STATISTICS

A Spectator Sport

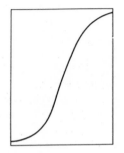

Richard M. Jaeger

SAGE Publications
Beverly Hills / London / New Delhi

STATISTICS

DEDICATION

To Judy, a never-ending source of inspiration and patience.

For information address:

SAGE Publications, Inc.
275 South Beverly Drive
Beverly Hills, California 90212

SAGE Publications India Pvt. Ltd.   SAGE Publications Ltd.
C-236 Defence Colony   28 Banner Street
New Delhi I 10 024, India   London EC1Y 8QE, England

Printed in the United States of America

Library of Congress Cataloging in Publication Data

Jaeger, Richard M.
Statistics, a spectator sport.

1. Statistics.   I. Title.
QA276.12.J33   1983   001.4'22   83-17740
ISBN 0-8039-2171-3
ISBN 0-8039-2172-1 (pbk.)

First printing, 1982, Edgepress

Second Printing 1983, Sage Publications

Table of Contents

Foreword

All professionals today—and arguably all citizens—need an understanding of statistics that goes well beyond the common knowledge of mean, median and mode. Few reports of medical advances, educational changes or social programs can be assessed as important for oneself or one's community without knowing (for example) what a standard deviation or a correlation coefficient signifies.

Now people come to understand how automobile engines work without learning how to build them. But statistics is usually taught by bringing out the formulas one would only need in order to calculate the statistics (and then only if one had lost one's $50 scientific calculator). For most people this approach does not prove enlightening.

It seemed time for a consumer's guide to statistics—but nevertheless one that was written by an experienced, practical statistician. My work for Nova University in providing seminars on evaluation, for hundreds of administrators, brought home the need for understanding statistics in this way. I was fortunate enough to be able to get Dick Jaeger to teach the statistics component and it seemed natural to ask him to go the extra step to a book. Thanks to Gerald Sroufe and Jim Johnson of Nova for the environment and early support; to the other statisticians on our Nova team—Alex Law and Bob Heath; to Nola Lewis and Bob Cooney of Edgepress for the production work; to Sara McCune of Sage for her confidence in us; and to Dick for his tireless, scholarly and pathbreaking effort.

Michael Scriven

Preface

It is virtually impossible to avoid statistics as a spectator sport. Statistics are everywhere—in baseball and football scores, in weather reports, in reports on stock and bond markets, and in descriptions of election prospects and outcomes. In addition, statistics form the bedrock of empirical research and evaluation in fields as diverse as education, the social sciences, and quantum physics. Certainly, if you hope to be an intelligent consumer of research in any of these fields, you must *understand* some statistics.

This book is designed for those who simply want to *understand* statistics, as they appear in the research and evaluation reports of educators, social and behavioral scientists, business and governmental reports, and so on. Those who want to learn how to *compute* statistics should seek some other text because, unlike most other books on statistics, this one has *absolutely no equations*. We have eliminated equations so as to avoid the frequent tendency of symbols to *obscure* the understanding of what is generally a very logical and straightforward subject area, especially for those with little mathematical expertise. By reading this book you will learn what statistics are, what they mean, and how they are used and interpreted. If you want to become an expert producer of statistics, you might start with this book, but you'll then have to move on to other texts that abound with x's, y's and a lot of Greek. We think you'll find them easier to handle once you've been through this one.

The first four chapters of the book are concerned with topics that fall under the general heading of "descriptive statistics." These include methods for organizing, tabulating, and graphing data; statistics that indicate the center of a distribution of scores (measures of central tendency); statistics that indicate the spread of a distribution of scores (measures of variability); and statistics that indicate the degree of relationship between two or more variables (measures of correlation). The fifth chapter is devoted to the basic measurement concepts of validity, reliability, and transformed score scales. Chapters 6 through 8 introduce

9

the fundamental concepts of inferential statistics: estimation and hypothesis testing, and form a conceptual framework for the remainder of the book. Chapters 9 through 14 describe applications of inferential statistics to a wide range of research and evaluation problems, beginning with inferences concerning a single population average, and ending with problems involving the prediction of one variable from a number of other variables (multiple regression analysis). The analysis of covariance and factor analysis are also described and illustrated in the final chapter.

To aid understanding and interpretation of statistics in the context of research and evaluation, we have provided detailed illustrations of each of the inferential statistical procedures discussed in Chapters 9 through 14. As illustrative material, since everyone has been through and continues to have a stake in the educational system, we have used excerpts from educational research and evaluation studies contained in the computerized files of the Educational Research Information Centers (ERIC). The reader will therefore encounter real research results in this text, and will learn how those results can be interpreted and used.

Although the title page bears the name of a single author, this book reflects the contributions of a great many people. It was first conceived as a joint work with Dr. Alexander Law of the California Department of Education. Dr. Law provided initial drafts of a number of chapters on descriptive statistics, and helped to shape the final text through many helpful and enjoyable conversations. Ms. Judy M. Jaeger and Dr. Donata Renfrow of Georgia State University edited major portions of the text, and made countless excellent suggestions on substance as well as editorial style. They deserve credit for resulting clarity, but certainly, none of the blame for remaining opacity. I am truly grateful for their help. Ms. Emily Barnes spent countless hours in the university library, digging for appropriate examples in the ERIC files. Her work was truly heroic. Participants in Nova University's National Ed. D. Program in Educational Leadership proofread major portions of an early draft of the book, and provided a number of helpful suggestions on layout and substance. Their assistance is gratefully acknowledged. It is unlikely that this book would have been written without the encouragement and helpful prodding of Dr. Michael Scriven of the University of San Francisco. The title and typesetting are his, and a detailed editing job in between.

Richard M. Jaeger
Greensboro, North Carolina
August, 1983

.

1.

Making Numbers Make Sense

Introduction and Overview

There are hundreds of statistics books in libraries and bookstores. When they were written, the author of every one thought his or hers was unique, and probably better than all the others, at least for some purposes. This book is different from the others in one obvious way, and that difference makes it better for some purposes.

Our statistics book doesn't have any equations in it. We don't know of any other statistics book that can meet that claim. If you want to know how to *compute* statistics, not having any equations is a serious limitation, and we recommend that you go to your bookstore or library and get someone else's book. We're particularly fond of Glass and Stanley's treatise on statistics for those working in the fields of education and psychology (1970). It is clear, correct, and contains a wealth of information. But if you just want to know how to *understand* statistics, as you will find them in research reports, journal articles, and evaluation reports, stick with us. Through diligent study of this book, paying particular attention to our real-life excerpts from research and evaluation reports, you'll learn a great deal about what statistics say and what they mean. And you can leave the computing to people who have the time, the need, and the inclination.

Many people find serious statistics scary or complicated. Actually, statistics are just the same friendly sort of numbers you've been dealing with all your life. They're things like baseball player's batting averages, the chances of rain tomorrow (when it's

11

expressed as a number), the odds on a Derby favorite, the total rainfall during the month of May, and the Dow-Jones stock average, to name just a few. The word "statistics" comes from the Latin root "status," which suggests that statistics should be useful in describing how things are.

Every discipline has its own set of concepts, terms and specialized language, and the field of statistics is no exception. Necessarily then, much of the early part of this book will consist of introducing concepts and giving their definitions in the context of problems and practical examples. We begin with four chapters on statistical concepts that are used to *describe*—the heart of **Descriptive Statistics.** We then spend a chapter on important things to know about **tests and measurements.** This is particularly important for people who work in education, psychology, or other social or behavioral sciences; measurement is especially tricky when people are being measured. The resulting numbers don't always mean what the measurer expects. It's useful to know where the pitfalls are, and in fact there are special statistics that describe the quality of measurement. Finally, if you stick with this book to the end, you'll find chapters on **Inferential Statistics**—a set of statistical procedures that allows a researcher to go beyond the group that's been measured, and make statements about the characteristics of a much larger group, just as the Gallup poll does.

You've probably noticed that we've printed some words in **boldface** type, and you've speculated that we think those are important terms. You're right. In fact, we think they're so important that we've included them in a glossary. Whenever you encounter a **boldface** term, throughout the remainder of the book, just look in the Glossary to find out what it means.

In the remainder of this chapter we begin our presentation of descriptive statistics. We start with a discussion of tabular arrangement of numbers (these numbers describe some facts about the world, so they are called **data**; one such fact is a "datum" and the plural is "data"), and we end by considering the pictorial display of data in **graphs.** In all of these procedures, as will be true throughout the book, we'll devote far more attention to what the tables and graphs *mean* than to the way they were *created*. Statisticians *create* them—you have to *understand* them.

Interpreting Data Tables

Table 1 contains the test scores earned by 40 students who completed an 85-item mathematics achievement test at the end of their ninth-grade school year. Their scores are not arranged in any particular order. We've listed the scores in eight different columns just so the table would fit nicely on the page. These data present just the kind of problem that descriptive statistics is designed to handle—how do you say something enlightening about a mass of numbers, how do you "boil it down"? (Remember, there might be 4,000 or 4,000,000 scores instead of 40.)

Table 1. Achievement Test Scores for a Class of Ninth-Grade Students

42	78	63	37	42	45	52	52
60	26	35	35	83	61	49	58
70	39	61	63	65	65	39	79
46	57	33	72	46	71	58	40
57	29	72	65	49	52	57	60

What can you tell about the performance of this group of students from this arbitrarily-arranged list of their scores? How easy is it to figure out what the highest score was? Can you readily find the lowest score? Did the scores tend to bunch up in a particular score area (say, between 50 and 70)? Overall, were there more high scores or more low scores? With enough effort, you could answer all of these questions. But you'd have to spend a great deal of time going over the numbers, and errors would be rather likely.

We've used the data in Table 1 to construct a number of tables and graphs that make it far easier to understand how the 40 students performed on the test. These tables and graphs are typical of those you'll find in countless research and evaluation reports, and in dozens of journal articles and statistics books.

Making sense of the data in Table 1 is greatly assisted by creating a new table in which the scores are arranged in order, from highest to lowest. Table 2 contains the results of this simple procedure, together with a useful description of the original scores in terms of their **ranks.** A table that lists scores from highest to lowest is called an **ordered array.** Table 2 is an ordered array that also contains an associated listing of score ranks.

**Table 2. Achievement Test Raw Scores
and Corresponding Ranks for a Class of Ninth-Grade Students**

Raw Score	Rank	Raw Score	Rank
83	1	57	20
79	2	52	23
78	3	52	23
72	4.5	52	23
72	4.5	49	25.5
71	6	49	25.5
70	7	46	27.5
65	9	46	27.5
65	9	45	29
65	9	42	30.5
63	11.5	42	30.5
63	11.5	40	32
61	13.5	39	33.5
61	13.5	39	33.5
60	15.5	37	35
60	15.5	35	36.5
58	17.5	35	36.5
58	17.5	33	38
57	20	29	39
57	20	26	40

Table 2 contains two sets of columns, each headed "Raw Score" and "Rank." The right-hand set of columns is merely a continuation of the left-hand set. We've used two sets of columns instead of one, just to get the table to fit nicely on a page.

The "Raw Score" columns list each of the 40 scores shown in Table 1. These are called **"Raw Scores"** because the data have not been manipulated or changed in any way. Each number represents a score, just as it was originally measured. We'll get to the ranks in just a moment.

Looking at the ordered listing of raw scores in Table 2, it is very easy to answer several of the questions we posed above. First, the highest-scoring student earned a score of 83. That's not bad on an 85-item test! Second, the score of the lowest-scoring student was only 26. You can tell immediately that the scores of the 40 students are spread out over a wide range. It seems that most of the students scored in the 40's, 50's and 60's, but it's not easy to tell whether the scores are bunched up within that range of values. Because each of the "Raw Score" columns lists 20 scores, and the score value 57 falls at the bottom of one and the top of the other, it appears that half the students earned scores above 57, while half scored below 57. Notice how much you can tell about the students' performance, just by looking at an ordered array!

The columns headed "Rank" are very useful for interpreting the performance of an individual student, compared to the performance of everyone else in the group. It should be obvious that the student who scored highest was assigned a rank of 1, the student who scored next highest was assigned a rank of 2, and so on. That simple rule works well until we get to the two values of 4.5 in the left-hand "Rank" column. Notice that two students earned scores of 72 on the mathematics test. It wouldn't make sense to say that one of those students ranked above the other, since they both earned the same score. So what ranks do we give them? Informally, we'd say "they were both 4th"; but statisticians use a slight variation of this. They first figure out what their ranks would have been, had they earned different scores. Suppose one had earned a score of 72 and the other a score of 71.5. Then the student who earned the 72 would have been assigned a rank of 4 and the student who earned the 71.5 would have been assigned a rank of 5. But since they earned the same score (72), we assign them both the average of the ranks they would have received, had their scores been different. In this case, the average of 4 and 5 is $(4+5)/2=4.5$. Then the student who earned a score of 71 was assigned a rank of 6, since five students earned scores that were above that student's 71.

In assigning ranks to the three students who earned scores of 65, we applied the same logic we used for the two students who earned scores of 72. Had the three students with scores of 65 all earned scores that differed slightly from each other, the highest-scoring student would have received a rank of 8, the next-highest a rank of 9, and the lowest-scoring a rank of 10. Since they all earned the same score, we give each of them the average of the ranks they would have received, had their scores been different. The average of 8, 9 and 10 is $(8+9+10)/3=27/3=9$. True, only 7 students did better than any of these, so you would informally say they were "all 8th." But they each have fewer students *below* them than someone who came in 8th without being tied. So we split the difference and average the ranks of those who tie.

Ranks allow you to interpret a score **normatively.** That's a fancy way of saying that ranks tell you how high a person's score is, *compared to the others* in the group that person was tested or measured with. Notice that the student who scored 40 on the test was assigned a rank of 32. Therefore, the student with a score of 40 (since no one else tied that student's score) is 32 students from the top; 31 of the 40 students in the group outscored that student. Compared to his/her classmates, the student with a score of 40 didn't make an impressive showing on the mathematics test.

When you look at a person's rank on a test or other measurement, you must be careful to remember that their rank is group-dependent. The same person might have a very different rank on the test, when compared with a different group. Compared to a group of 39 mathematics wizards, the student with a rank of 1 in Table 2 might have received a rank of 20 or 30. This would have happened if most of the mathematics wizards had earned perfect scores on the 85-item test.

Now let's change topics. Consider one of the important truisms of statistics: The more numbers a person has to think about at the same time, the more difficult it is to figure out what those numbers mean. This argues strongly for summarizing data as much as possible. The tables we'll consider next are motivated by the desire for summarization.

One way to summarize a group of scores is by eliminating those that are redundant. Table 3 is an example of something called a **frequency distribution,** in which this is done.

Notice that Table 3 consists of two columns, one headed "Score" and the other headed "Frequency." The "Score" column contains a list of each of the score values that appeared in the original data in Table 1. Even though there were two scores of 72 in the original data, the score value 72 is listed only once. This is what we mean by eliminating redundant scores. To keep track of how many times each score value appeared in the original data, we show this number in the column headed "Frequency." Since there was only one score of 83 in the original data, its frequency is 1. Likewise, since they appeared only once, scores of 79 and 78 receive frequency values of 1. The score value 72 receives a

**Table 3. Frequency Distribution of Achievement Test Scores
for a Class of Ninth-Grade Students**

Score	Frequency
83	1
79	1
78	1
72	2
71	1
70	1
65	3
63	2
61	2
60	2
58	2
57	3
52	3
49	2
46	2
45	1
42	2
40	1
39	2
37	1
35	2
33	1
29	1
26	1

frequency of 2 because there were two 72's in the original data.

Instead of having 40 scores to consider, as was true in Table 2, in Table 3 we have only 24 unique score values and their associated frequencies. This kind of summarization has advantages beyond a mere reduction in the number of scores to be considered. From Table 3 you can get an impression of the "shape" of the distribution of students' scores. Notice that most of the scores with frequencies larger than 1 fall in the middle of the distribution. In contrast, almost all of the very high scores and very low scores have frequency values of 1. From this you can conclude that the students' scores tend to bunch up in the middle of the distribution. Looking at the data in Table 3, it is also easy to tell which score values occur most often; the values 65, 57, and 52 all have frequencies of 3. All of the other score values occur less frequently.

Additional summarization of the scores in Table 2 is provided by the **grouped frequency distribution** shown in Table 4. "Grouping" further reduces the number of separate lines on the table, as you can see; Table 3, by contrast, is an **ungrouped frequency distribution.**

**Table 4. Grouped Frequency Distribution of
Achievement Test Scores
for a Class of Ninth-Grade Students**

Score Class	Frequency
80-84	1
75-79	2
70-74	4
65-69	3
60-64	6
55-59	5
50-54	3
45-49	5
40-44	3
35-39	5
30-34	1
25-29	2

To construct Table 4, we've grouped the original scores into categories of "Score Classes." We then counted the number of scores that fell into each score class, and listed these counts in the column headed "Frequency." Thus two score values (one 26 and one 29) fell into the score class "25 to 29"; one score (the score of 33) fell into the score class 30 to 34; etc.

By constructing Table 4, we have further reduced the number of score values that must be considered when interpreting the test performance of the 40 students. Instead of the 40 scores of Table 2 or the 24 unique score values of Table 3, we need only consider 12 score classes and their associated frequencies. Our goal of summarization is well served by this reduction.

From Table 4 it is easy to get a picture of the overall test performance of the group of 40 students. For example, notice that five students earned scores between 55 and 59, and that an additional six students earned scores between 60 and 64. Adding these frequencies together, we can say that 11 of the 40 students (better than a fourth of them) earned test scores between 55 and 64. Doing more of the same kind of data sleuthing, we can add the frequencies two, one and five associated with the lowest score classes 25 to 29, 30 to 34, and 35 to 39, respectively. From this total frequency of 8, we can say that 8 out of 40 (or 20 percent) of the students earned test scores that were below 39 on the 85-item test. Depending on our expectations for student performance, this might be a distressing finding.

Although grouped frequency distributions afford the advantage of summarization, they have an associated disadvantage. The price of summarization is losing the detailed information about the original data that is provided by an ungrouped frequency distribution. For example, had we begun with Table 4, and not had Tables 1 through 3 available, we could have concluded that two students earned test scores *somewhere* between 25 and 29, one student scored somewhere between 30 and 34, etc. However, we would not have known whether both of the students who scored between 25 and 29 earned scores of 25, or whether one scored 26 and the other 28, or whether both scored 29. Likewise, we would not have known whether the one student with a score in the 30 to 34 class scored at the top or the bottom of that class. This detailed information is sacrificed in order to summarize a set of data in a grouped frequency distribution. For some purposes the sacrifice is serious; for others, it is not serious.

Rules for constructing grouped frequency distributions are arbitrary, and vary from one statistics book to another. One author suggests that a grouped frequency distribution should have between 10 and 20 score classes. Another suggests that 12 to 15 score classes are appropriate. All of these rules are attempts to compromise between the desire to summarize data and the desire to retain as much detailed information as possible. The greater the number of score classes, the more detailed information about the original data is retained. The smaller the number of score classes, the more the data have been simplified.

Two additional data displays that are used quite often in research and evaluation reports are illustrated in Table 5. This table combines the grouped frequency distribution of Table 4 together with a **cumulative frequency distribution,** and a **cumulative percent frequency distribution;** long names, but easy to understand.

Table 5. Grouped Frequency Distribution, Cumulative Frequency Distribution and Cumulative Percent Frequency Distribution of Achievement Test Scores for a Class of Ninth-Grade Students

Score Class	Frequency	Cumulative Frequency	Cumulative Percent Frequency
80-84	1	40	100
75-79	2	39	97.5
70-74	4	37	92.5
65-69	3	33	82.5
60-64	6	30	75
55-59	5	24	60
50-54	3	19	47.5
45-49	5	16	40
40-44	3	11	27.5
35-39	5	8	20
30-34	1	3	7.5
25-29	2	2	5

The "cumulative frequency distribution" part of this consists of the columns headed "Score Class" and "Cumulative Frequency." Let's first consider where the numbers in the Cumulative Frequency column came from, and then we'll discuss the way they can be used.

Notice that the cumulative frequency associated with the *bottom* score class 25-29 is equal to the frequency of that score class, namely 2. The frequency of 2 means that two students scored somewhere between 25 and 29. The reason we say that the students scored in this interval is because, in scoring the mathematics achievement test, we count the number of items a student answered correctly. The only possible scores are, therefore, zero, 1, 2, etc., up to a maximum score of 85.

Even if we were able to measure students' mathematics achievement more precisely than this, say as precisely as we wanted, we could still—for convenience—report a student's score to the nearest whole number, by "rounding" their actual scores in the traditional way. This means that any student who had a mathematics achievement level between 28.5 and 29.4999999 (for example) would have a reported whole-number score of 29. Any student who had a mathematics achievement level between 24.5 and 25.4999999 would have a reported whole-number score of 25. Normally, numbers ending in fractional values of 0.5 or above are rounded up, and numbers ending in fractional values below 0.5 are rounded down. Reasoning in this way, we could say that the score class 25 to 29 really represents mathematics achievement levels between 24.5 and 29.4999999. Because it's a nuisance to keep writing a long number like 29.499999, it is traditional to say that the score class 25-29 really represents values between 24.5 and 29.5. These values are called the **real limits** of the score class 25-29. The value 24.5 is called the **lower real limit**, and the value 29.5 is called the **upper real limit.** Score classes are really a way to "round-off" scores into even cruder categories than whole numbers.

Now we can give an exact definition of the numbers in the Cumulative Frequency column of Table 5. Beginning with the value 2 associated with the 25-29 score class, the numbers in this column correspond to the numbers of students who earned scores at or below the upper real limit of each score class. Using this explanation, we can see that the table says two students earned scores at or below 29.5, three students earned scores at or below 34.5, eight students earned scores at or below 39.5, 11 students earned scores at or below 44.5, etc. Finally, the numbers in the Cumulative Frequency column indicate that all 40 students earned scores at or below 84.5, since the largest cumulative frequency is 40.

Where did the numbers in the Cumulative Frequency column come from? Let's illustrate. Looking at the previous table, you can see that two students scored in the 25-29 score class, and one student scored in the 30-34 score class; we merely add the two to the one to conclude that three students scored at or below the upper real limit of the 30-34 score class. The other numbers in the Cumulative Frequency column (called "cumulative frequencies") were found through similar logic.

The cumulative frequency distribution provides another useful way to depict the overall test performance of the group. Recall that the mathematics achievement test consisted of 85 items. Just by selecting a few values from the cumulative frequency distribution, we can get a pretty clear picture of the group's performance on this test. Notice that 11 of the 40 students earned scores of 44.5 or less. This means that more than a fourth of the students missed almost half the items on the test. Three-fourths of the students (30 out of the 40) earned scores at or below 64.5. This means that three-fourths of the students missed at least 20 of the 85 test items. Other cumulative frequencies could be interpreted similarly, but these two should give you an idea of the usefulness of the cumulative frequency distribution in summarizing a group of scores.

The right-most column in Table 5 lists "Cumulative Percent Frequency." Together with the "Score Class" column, the Cumulative Percent Frequency column forms a cumulative percent frequency distribution—another very useful data display. The cumulative percent frequencies were calculated simply by converting the cumulative frequencies to percentages of the total number of students in the group. For example, notice that the 60-64 score class has a cumulative frequency of 30 and a cumulative percent frequency of 75. The 75 was calculated as follows: The cumulative frequency of 30 was divided by the group size of 40: 30/40=0.75. The resulting proportion (0.75) was then multiplied by 100 to convert it to a percent (75 percent). From this figure we can see directly that 75 percent of the students in the group earned test scores at or below 64.5. Likewise, 20 percent of the group earned scores at or below 39.5; 40 percent of the group earned scores at or below 49.5; 60 percent of the group earned scores at or below 59.5; and 92.5 percent of the group earned scores at or below 74.5. These statistics provide a very clear description of the test performance of the group of 40 students. In fact, you can probably get a better idea of the performance of the group from these five numbers than you could from the entire set of 40 scores shown in Tables 1 and 2. Typically, effective summarization of data enhances understanding.

Grouped frequency distributions are often useful in comparing the performances of several groups. For example, Table 6 illustrates the distributions of scores of students in three schools or instructional programs, labeled "A," "B," and "C," as well as the distribution of scores for all three groups of students combined.

Notice how readily you can see that Group B scored highest, and that Group C was midway between Groups B and A. This table makes it easy to address the question, "Did one of these programs or schools do a better job—or have better students—than another?" In addition, from the combined distribution you can see a concentration of scores between 60 and 64, and another large concentration between 70 and 79, affording a picture of the aggregated performance distribution of all three groups.

Table 6. Grouped Frequency Distribution of Scores for Students in Three Schools or Programs

| Score | Number of scores recorded by school or program | | | |
	A	B	C	Combined
95-99		1		1
90-91		2		2
85-89		5	1	6
80-84		9	8	17
75-79	1	11	12	24
70-74	6	5	14	25
65-69	9	2	3	14
55-59	10		1	11
50-54	8		2	10
45-49	3			3
40-44	1			1

A Picture is Worth A Thousand Words— and at Least Ten Thousand Numbers

Tables are invaluable for communicating detailed information about a set of scores, but many people find it difficult to get the "big picture" from a table. To get an overall impression of the shape of a score distribution, or to get a quick idea of how several groups compare, there is no substitute for a **graph**. Graphs are irreplaceable tools in descriptive statistics. Personal computers and business computers are increasingly being used to display "graphics" (that is, graphs which summarize data) because graphs often are much easier to understand than even the best tables.

There are so many different types of graphs that it would be possible to devote an entire book to the subject. In fact, Mary Spear did essentially that in her 1952 volume entitled *Charting Statistics*. We will confine our discussion to just a few of the more popular and useful types of graphs that you are likely to encounter in your study of research and evaluation reports. If you want more information, check with Spear.

The **frequency polygon** shown in Figure 1 is a picture of the grouped frequency distribution in Table 4. When we drew the figure, we used the data in Table 4 as follows: First, we calculated the score value that would be right in the middle of each score class. For example, 27 is in the middle of the 25-29 score class; 32 is in the middle of the 30-34 score class; 37 is in the middle of the 35-39 score class, and so on. These middle score values are called the **midpoints** of the score classes. Next, we assumed that the score earned by every student who scored anywhere in a given score class fell at the midpoint of that score class. We then plotted the midpoints of each score class on the horizontal scale (called the "x-axis" or **abscissa**) of Figure 1, and the frequency values associated with each score class on the vertical scale (called the "y-axis" or **ordinate**) of Figure 1. Once we had constructed the horizontal and vertical scales, we plotted a point on the graph for each score class. Notice that there is a point just above the score value of 27 (the midpoint of the 25-29 score class) and just to the right of a frequency of 2. Similarly, there is a point just above the score value of 32 (the midpoint of the 30-34 score class) and to the right of a frequency of 1. Use the data in Table 4 to find each of the other points plotted in Figure 1. Once the points were plotted, we connected adjacent points with straight lines, and then tied the resulting curve down at the ends by plotting one point at a score value of 22 and a frequency of zero, and another at a score value of 87 and a frequency of zero. The score values of these endpoints are equal to the midpoints of the score classes that would be just below, and just above, the lowest and highest score classes in Table 4. Since no students scored in these score classes, their frequencies would be zero.

Notice that the horizontal scale in Figure 1 is broken between zero and 22. The figure contains two short diagonal lines to alert you to this fact. We broke the horizontal scale because the distance between zero and 22 is not the same as the distance between 22 and 27, or the distance between any of the other plotted

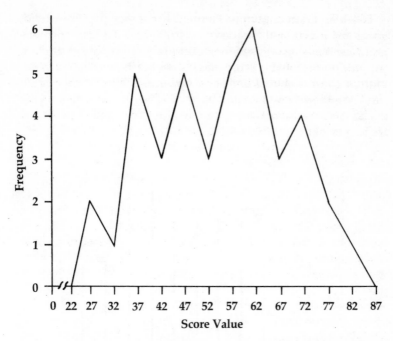

Figure 1. Frequency Polygon of Achievement Test Scores for a Class of Ninth-Grade Students

midpoints. In other words, the curve really doesn't start at zero. Since most of us expect the horizontal and vertical scales of a graph to cross at the zero-point, we must be warned, so we won't be misled, whenever a graph does not begin at zero.

Figure 1 does an excellent job of providing an overall impression of the test performance of the 40 ninth-grade students. We can see at a glance that their scores range from the low 20's to the high 80's. It is easy to see that more of them scored in a score class with a midpoint of 62 than in any other score class. The wide base of the peak that occurs at 62 tells us that many of the students scored in the 60's. The two smaller peaks below the highest one tell us that a number of the students scored in the high 30's and the high 40's as well. We can also see that the students' scores are spread out quite a bit across the middle of the distribution. No single score class has an extremely high frequency, compared to all of the others.

The **histogram** shown in Figure 2 is another popular way of graphing a grouped frequency distribution. To construct this graph we have again used the data in Table 4. Instead of plotting the midpoints of the score classes on the horizontal axis, we plotted the lower real limit and upper real limit of each score class. We then drew a vertical bar for each score class, making the bar as wide as the real limits of the score class, and as tall as the frequency of the score class.

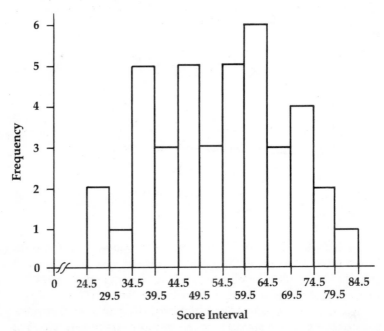

Figure 2. **Histogram of Achievement Test Scores for a Class of Ninth-Grade Students**

A frequency polygon and a histogram tell the same kind of story. Whether a writer uses one or the other to depict a grouped frequency distribution is strictly a matter of aesthetic taste. If a single graph is used to show two or more distributions (such as a distribution of scores for boys and the corresponding distribution for girls), a frequency polygon is often easier to read than a histogram. When one histogram is placed on top of another, the

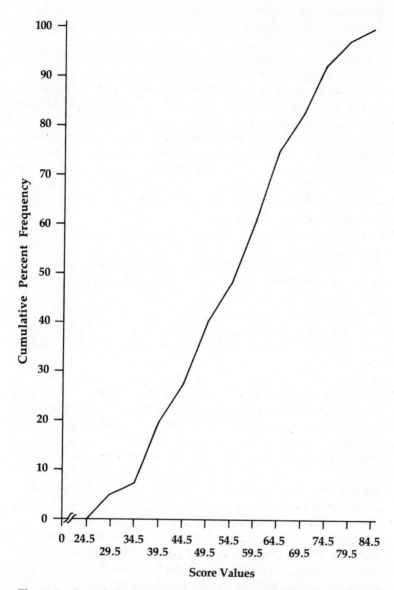

Figure 3. Cumulative Percent Frequency Polygon of Achievement Test Scores for a Class of Ninth-Grade Students

graph gets cluttered and confusing, though you can sometimes use color or shading or perspective to avoid confusion. (See Fig. 10 below.)

Look at the histogram in Figure 2, and figure out how many students scored in the score class with real limits of 69.5 and 74.5. Did fewer students score in the interval with limits 39.5 and 44.5? How many scored in the interval with limits 34.5 and 39.5? Your answers should be 4, yes, and 5.

The **cumulative percent frequency polygon** shown in Figure 3 is a picture of the cumulative percent frequency distribution contained in Table 5. Although this graph doesn't give an immediate visual impression of the score distribution, it is among the most useful types of graphs in descriptive statistics. So what's it good for, you ask? Let's see.

Suppose you wanted to know the point on the score scale that would divide the 40 students exactly in half, in the sense that 20 of them earned scores above that point and 20 of them earned scores below. That score value is called the 50th **percentile**. Any conventional statistics book worth its salt will have a formula for calculating the 50th percentile. But for most practical purposes, you don't have to know the 50th percentile exactly, an approximation will do just fine. And you can use Figure 3 to get it, as we'll show you.

To find the 50th percentile, just read up the vertical scale of Figure 3 (the Cumulative Percent Frequency scale) to 50. At that point, draw a horizontal line, parallel to the x-axis, across to the curve. When you're finished with this step, your graph should look like Figure 4. Next, at the point your horizontal line crosses the curve, drop a vertical line, parallel to the y-axis, down to the horizontal scale. When you're finished with this step, your graph should look like Figure 5. The point where your vertical line crosses the x-axis is the 50th percentile. In this case, the 50th percentile is about 55.

You can find other percentiles just as readily from Figure 3. For example, to find the point on the score value that separates the lowest-scoring 25 percent of the students from the highest-scoring 75 percent (called the 25th percentile), just follow the steps for finding the 50th percentile with one exception. Instead of reading up the Cumulative Percent Frequency scale to 50 in the first step, just go up to 25 (half way between 20 and 30). Draw your horizon-

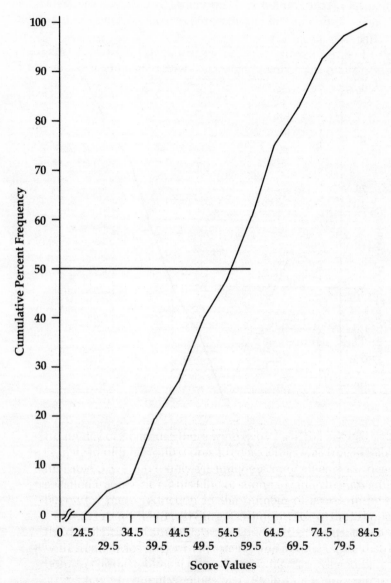

Figure 4. Cumulative Percent Frequency Polygon of Achievement Test Scores for a Class of Ninth-Grade Students

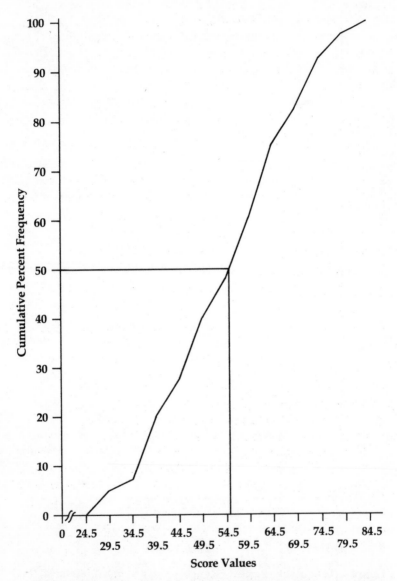

Figure 5. Cumulative Percent Frequency Polygon of Achievement Test Scores for a Class of Ninth-Grade Students

tal line across to the curve at that point, and follow the rest of the steps in order. You should find that the 25th percentile is about 43. The 25th percentile is used so often in statistics that it has a special name. It is called the **first quartile,** and is often denoted by Q_1.

To make sure you can use Figure 3 to find percentiles, try to determine the 75th percentile. This is another frequently-used statistic, that goes under the alias of the **third quartile** and is often denoted by Q_3. It is the point on the score scale that separates the lowest-scoring 75 percent of the students from the highest-scoring 25 percent. The 75th percentile is equal to 64.5.

If you wanted to describe the score distribution of the 40 students in very great detail, you could use Figure 3 to find 99 percentiles altogether. One percent of the students would have score values at or below the first percentile; two percent would have score values at or below the second percentile; and so on, up to the 99th percentile. It is rarely necessary to know all 99 of a distribution's percentiles, but it's nice to know they're around in case a complete description is needed.

Although you might be tired of Figure 3 by now, we've not yet exhausted its wonders. We can use it to generate yet another set of useful statistics. In an early part of this chapter, we described the use of ranks for interpreting a student's score in comparison with the scores of the other students. **Percentile ranks** are another set of statistics that perform a similar function, and we can determine approximate percentile ranks using Figure 3.

A student's percentile rank is equal to the percentage of students with scores that are less than or equal to his or her score. For example, a student who scored 54 on the mathematics achievement test would have a percentile rank of 47 (denoted P.R.=47). This means that 47 percent of the 40 students earned scores of 54 or less.

We determined the percentile rank of the student who scored 54 by using Figure 3 as follows: First, we found the score value 54 on the horizontal axis of the figure. From that point, we drew a vertical line, parallel to the y-axis, up to the curve. At that point, we had a graph that looked like the one shown in Figure 6.

Next, at the point where our vertical line crossed the curve, we drew a horizontal line, parallel to the x-axis, over to the Cumulative Percent Frequency scale. This gave us a graph like the one

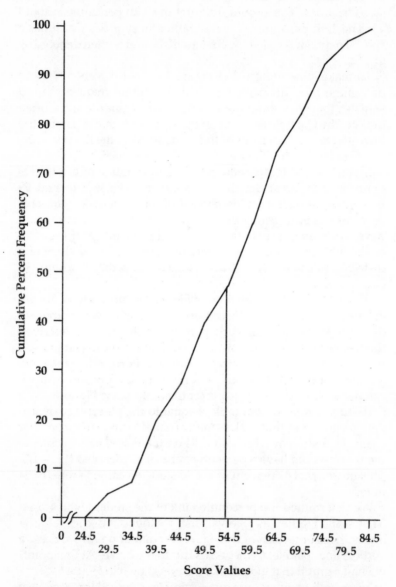

Figure 6. Cumulative Percent Frequency Polygon of Achievement Test Scores for a Class of Ninth-Grade Students

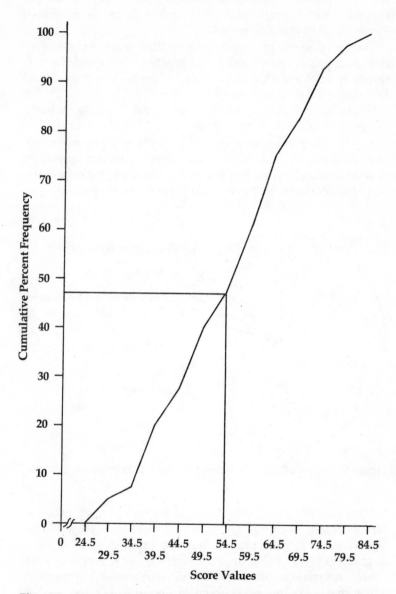

Figure 7. Cumulative Percent Frequency Polygon of Achievement Test Scores for a Class of Ninth-Grade Students

shown in Figure 7. Our horizontal line crossed the Cumulative Percent Frequency scale part way between 40 and 50, so we read the percentile rank of 54 to be 47.

You can find the percentile rank of other scores by following the same procedure we used. For practice, try to find the percentile rank of a student who scored 66 on the test. You should find that the P.R. of 66 is about 77.

Unlike most statistics, which describe the characteristics of a group of individuals or their scores, percentile ranks are used to interpret the performance of one individual. Percentile ranks provide the basis for a *normative* (i.e. comparative) interpretation of an individual's score, just like ranks. Percentile ranks have the added advantage of being on a 100-point percentage scale.

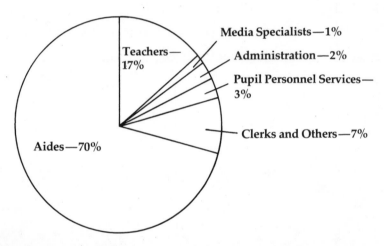

Figure 8. Types of Personnel Funded by ESEA Title I Projects, 1975-76.

Some Fancier Graphs The types of graphs we've described so far are among the more common and most widely used in research and evaluation reports. But you'll find many other kinds of graphs in your wanderings through the research of your field, as well as variations on the themes we've discussed. For example, a **pie chart**, like the one shown in Figure 8, is often used to illustrate a breakdown of some quantity, like income, expenditures, or personnel, across categories. A pie chart shows quite clearly which category is getting the largest slice. In the case of Figure 8, we can conclude without question that the vast majority of personnel funded by Title I of the Elementary and Secondary Education Act (ESEA) in 1975–76 were classified as Aides. The Aides slice is 70% of the entire pie!

Sometimes bar charts (a more common name for histograms) are presented horizontally instead of vertically. The graph in Figure 9 is an example of a **horizontal bar-chart**. Notice that this histogram shows the distribution of students who participated in the ESEA Title I program in 1975–76 by grade level. Since the grade levels occur in discrete units (kindergarten, first grade, second grade, etc.), there really are no *score* classes and it is not necessary to plot real limits of score classes on one of the scales.

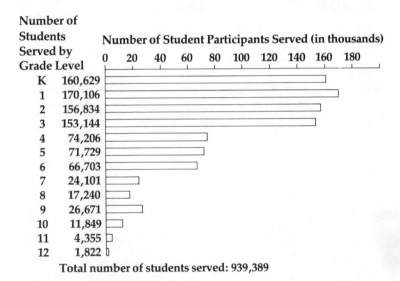

Figure 9. Number of Student Participants in ESEA Title I, by Grade Level, 1975-76.

To show a graphical comparison of two score distributions, a graph similar to the **double-barred histogram** in Figure 10 is often useful. This graph shows the average scores earned by students in grades one, two, and three, who participated in an early childhood education program during the 1976–77 school year.

The average scores earned by these students at the beginning of the school year (Pretest) and at the end of the school year (Post-test) can be compared quite readily. It is also easy to see how the average scores compare to the national average of 50.

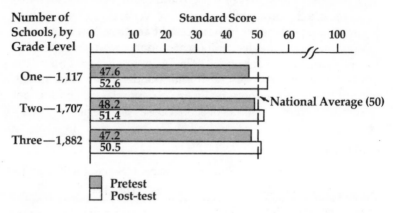

Figure 10. **Pretest and Post-test Standard Scores in Mathematics Achievement, by Grade Level, for Schools Participating in Early Childhood Education Programs, 1976-77**

Describing the Shapes of Curves   Statisticians and researchers have developed a set of terms to describe the shapes of frequency polygons and other curves. Their specialized language is a form of jargon, but it saves many extraneous words, once you know the language. Compare the following descriptions of the same frequency polygon: (1) the frequency distribution was very tall and skinny; it had one great big peak but the rest of it was much flatter toward the higher end of the score scale. (2) The frequency distribution was leptokurtic, unimodal, and positively skewed. These two descriptions are identical. Using a few technical terms, the second is far more concise and, once you know the language, more communicative.

Over the centuries, many frequency distributions describing real data have been found that have one particular shape. They have one peak in the middle of the distribution, and their frequencies get progressively smaller, as one moves away from the middle of the distribution. Furthermore, the shape of these distributions *above* their middle value is a mirror image of their shape *below* the middle.

In the early 1700's a mathematician named Abraham De Moivre developed the mathematical equation for a curve that approximates these distributions very well. This **normal distribution,** when graphed, produces a bell-shaped **normal curve** like the one shown in Figure 11. Through the years, statisticians have adopted the normal curve as a kind of benchmark, to be used as a reference when describing other frequency polygons.

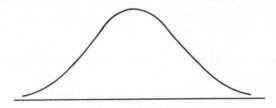

Figure 11. The Normal Curve

The normal distribution is extremely useful, not only for approximating the shape of many frequency distributions, but for solving a variety of problems in inferential statistics as well. In fact, it is not an exaggeration to say that the field of statistics would still be in its infancy, had the normal distribution never been derived. We will return to the wonders of the normal distribution repeatedly, in later chapters. For the moment, just notice the friendly, familiar shape of the normal curve. It is neither too tall and skinny, nor too short and fat. In fact, it is regarded as a standard of moderation and middle-of-the-roadness in terms of the height and width of its peak. The normal curve is therefore labeled **mesokurtic.** That's a fancy Latin-root word that comes from "meso" meaning "middle" and "kurtic" meaning "peakedness." So the normal distribution is, as we say in the South, middling in its peakedness.

A frequency distribution that is taller and skinnier than the normal distribution, like the one shown in Figure 12, is called **leptokurtic.** To remember that label, just think of the distribution as "leaping up" in the middle. If you were to give an achievement test to a large sample of "C students" selected from a high school,

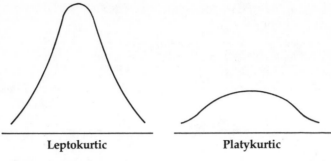

Leptokurtic   Platykurtic

Figure 12. A More-than-   **Figure 13. A Flattened Curve**
normally Peaked Curve

their distribution of test scores would probably be leptokurtic, like the curve in Figure 12. Most of the students would score close to the middle of the distribution. There would be few, if any, students with very high or very low test scores.

A frequency polygon that is flatter, and more spread out, than a normal distribution is labeled **platykurtic.** If you were to select a 500-student sample from a high school, consisting of 100 "A students," 100 "B students," 100 "C students," 100 "D students" and 100 "F students," a distribution of their achievement test scores would look like the platykurtic curve shown in Figure 13. You would expect the test scores of these students to cover a wide range of values, with approximately equal numbers of students scoring in the middle of the distribution and in the high and low score ranges.

To associate the term "platykurtic" with a relatively flat frequency polygon, just think of a duck-billed platypus. Agreed, you don't see many of those little critters wandering around the streets these days, but most people have heard of them, and they are wide and flat. If you can find a more-familiar mnemonic device, by all means use it!

All of the frequency polygons we've described in this section share some important characteristics. Each of them has only one peak. And if we were to draw a vertical line right through the middle of each curve, its shape to the left of that vertical line would be a mirror image of its shape to the right of that vertical line. In other words, the left half of each curve would fall right on top of the right half, if the curves were folded around a vertical line in the middle. Both of these characteristics have statistical names. A frequency polygon that has only one peak is called **unimodal**. A frequency polygon that has mirror-image shapes to the left and right of its middle is called **symmetric**. Although many frequency distributions are both symmetric and unimodal, there are exceptions.

The frequency polygon shown in Figure 14 has two peaks. It is therefore called **bi-modal**. If you were to graph the frequency distribution of heights in inches, for a sample consisting of 100 male basketball players and 100 male jockeys, you would probably have a bi-modal curve like the one shown in Figure 14. If a frequency polygon has more than two peaks, it is called **multimodal**.

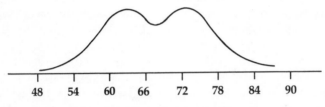

Figure 14. **A Bi-modal Distribution Curve**

A frequency polygon that is shaped so that the curve to the left of its middle is *not* a mirror image of the curve to the right of its middle is called **skewed**. Many frequency distributions result in skewed frequency polygons. For example, suppose you were to collect data on annual salaries from people who happened to pass you on a downtown street corner in a randomly-chosen American city. Most of the people who passed your corner would probably have low to moderate annual salaries. But there would be a few bank presidents and physicians, whose salaries would be well above those of the crowd. If you were to plot a frequency polygon of these salary data, it would probably look like the curve in Figure 15.

Notice that the salaries with high frequencies are in the left-hand end of the polygon (at the low end of the score scale) and that salaries at the high end of the scale have relatively low frequencies. This kind of frequency polygon is said to be **posi-**

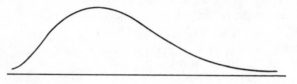

Figure 15. A Positively Skewed Curve

tively skewed because the distribution is stretched out toward the high (or positive) end of the scale. (If you think of it as a diagram of a hand with one finger sticking out, the finger points in the direction of positive values on the x-axis.)

As you might expect, the converse of a positively skewed frequency polygon is called **negatively skewed**. A negatively skewed polygon is illustrated in Figure 16. If you were to sample 200 average adult males, and ask them each to run a timed 100-yard dash, and then plot a distribution of their running times, you'd probably find a negatively skewed curve like the one in Figure 16. Think of time needed to run 100 yards as being plotted on the x-axis and frequency being plotted on the y-axis. Relatively few men would be in superb physical shape, and would run the 100-yard-dash in a short time (10 to 11 seconds). However, most of the men would take a relatively long time to run 100 yards (20 to 25 seconds), so that the frequencies associated with these longer running times would be relatively high.

Figure 16. A Negatively Skewed Curve

Well, enough of new terms for describing curves. Now when you read a research report that describes a distribution of scores as being unimodal, platykurtic, and symmetric, you can conjure up a mental image of the shape of its frequency polygon.

Problems—Chapter 1

1.1 A standardized achievement test in social studies was used in an evaluation of an innovative curriculum for sixth-graders. The 50-item test was administered to 100 participating students at the end of a one-year program, and the following cumulative frequency distribution was constructed using the resulting data:

Score Interval	Cumulative Frequency
46-50	100
41-45	96
36-40	91
31-35	75
26-30	42
21-25	21
16-20	10
11-15	5
6-10	2
1- 5	0

Which of the following statements is supported by the data shown in the table?

  a. None of the students scored above 46 on the test.
  b. The majority of students scored below 36.
  c. The majority of students scored above 35.
  d. None of the students scored below 10.

1.2 Suppose that the distribution of employee salaries in your administrative unit looked like the curve shown below. If you were trying to describe this distribution to a fellow administrator in a phone conversation, would you say that it was:

  a. Symmetric and unimodal.
  b. Negatively skewed and bi-modal.
  c. Positively skewed and bi-modal.
  d. Positively skewed and unimodal.

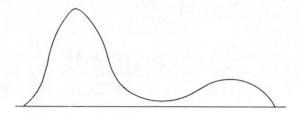

1.3 The distribution of professional employees per school in the River Junction Unified School System is shown by the following histogram. What category of number of professional employees has the highest frequency of schools? _____ How many schools fall into this category? _____

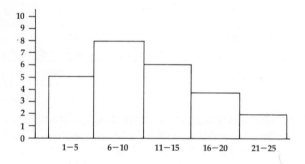

1.4 Look at the double-barred histogram shown in Figure 10 of Chapter 1. If you were interpreting the data reported in that figure, which of the following would be correct statements? (Circle the letter of *each* statement that is true.)

   a. More schools participated in the early childhood education program at the third-grade level than at the first-grade level.

   b. More schools participated in the early childhood education program at the first-grade level than at the second-grade level.

   c. The largest percentage of students with achievement above the National average on the post-test occurred at the first-grade level.

   d. The greatest apparent gain in achievement, from pretest to post-test, occurred at the first-grade level.

2.

Concepts Of Central Tendency

Where is the Center?

As we noted in Chapter 1, one of the principal functions of descriptive statistics is summarization of data. We often use a set of data to compute other numbers, called **statistics,** that convey the essential information in the original scores.

One of the most elementary types of statistics indicates the location of the center of a score distribution. In many statistics texts, statistics that identify the center or middle of a set of scores are called **measures of central tendency.** In this chapter we'll define several measures of central tendency, examine their properties and assumptions, and try to give you a sense of when each of them should be used. With some sets of data, various measures of central tendency will give totally different indications of the center of a score distribution. It is important to learn when each measure of central tendency is appropriate, so you won't be misled by an author who has used the wrong one.

How to Define the Center

The center of a score distribution can be defined in many different ways. However, the most commonly used measures of central tendency are the **mode,** the **median,** and the **mean.**

When used with a listing of scores, an ordered array, or an ungrouped frequency distribution (see Tables 1, 2 and 3 of Chapter 1), the mode is defined as the most frequently occurring score. In a grouped frequency distribution, the score class that has the highest frequency (biggest number) associated with it is called the **modal interval.** The *mode* is the score value that is right in the middle (the midpoint) of the modal interval. Many statistics texts use the symbol "Mo." to denote the mode. However, various authors use different symbols for the same statistics. A committee on statistical notation has existed within the American Statistical Association for over 30 years, and it's no closer to achieving consistent statistical notation than is the United Nations to bringing about world peace.

Without using the term, we described the *median* of a score distribution in Chapter 1. The median is another name for the 50th percentile. Recall that the 50th percentile is the score value that divides a distribution exactly in half, in the sense that 50 percent of the scores in the distribution fall at or below that value, and 50 percent of the scores fall above that value. Another name for the median is the **second quartile.** Just remember that the median, the 50th percentile, and the second quartile are all aliases for the same statistic.

Most statistics texts and research reports abbreviate the median by calling it "Md." or "Mdn." The second quartile is often denoted by Q_2.

The most widely used measure of central tendency is the arithmetic *mean*. You have been computing means for years, and you probably just called them averages. If you are working with a set of scores, their arithmetic mean is computed by simply adding the scores together and dividing by the number of scores. In most research and evaluation reports, you will find the mean denoted by the letter X with a line over it: $\overline{X}$. This is read as "X bar."

Which Measures When?

Three factors should be considered when you try to decide whether an author has used the right measure of central tendency: (1) the level of measurement of the data, (2) the shape of the score distribution, and (3) the stability of the measure of central tendency. We'll discuss these factors in order, since that's the way we'd consider them.

Level of Measurement. To introduce level of measurement issues, which relate to the *kind* of measurement involved, we'd like to give you some information and then ask you some questions.

Consider the following data, concerning four fictitious persons:

Joe	Sally	Mary	Bob
60	70	110	120

The four numbers written below the names are IQ-test scores. To avoid pages of interesting, but diverting, discussion, let's agree that we'll *assume* IQ-tests measure intelligence with perfect accuracy.

Given these assumptions, here are our questions: We ask that you respond affirmatively only to *one* of them.

(1) Notice that Bob's IQ-test score is twice as large as Joe's. Given this fact, are you willing to state that Bob is *twice as intelligent* as Joe?

(2) Notice that the difference in IQ-score between Sally and Joe is 10 points. The difference in IQ-score between Bob and Mary is also 10 points. Given this fact, are you willing to state that the *difference in intelligence* between Sally and Joe is the same as the *difference in intelligence* between Bob and Mary? (After all, 10 points is 10 points!)

(3) Notice that Bob has the highest IQ-test score, Mary is next, Sally's score is third from the top, and Joe's score is lowest. Given this fact, are you willing to state that Bob is *most intelligent,* Mary is second in *intelligence,* Sally is next, and Joe is *lowest in intelligence?* (That is, will you rank them by their scores?)

(4) Notice that Joe, Sally, Mary and Bob all have different IQ-test scores. Given this fact, are you willing to state that these four people *differ in their intelligence,* but not state anything about the relative levels of their intelligence?

These four questions correspond to four different assumptions about the relationship between the results of measurement (the score on the IQ-test), and the variable that is being measured (intelligence). Since various statistics incorporate assumptions like these, it is important to understand them.

Had you been willing to state that Bob was twice as intelligent as Joe because his IQ-test was twice as large as Joe's, you would have been assuming that the IQ test measured intelligence at the **ratio level** of measurement. This is the strongest measurement assumption you could make. With ratio-level measurement, a score that is twice as large as another corresponds to twice as much of the variable being measured as does the other score. In addition, a score of zero indicates that the amount of the variable is zero. In other words, if an IQ test measured intelligence at the ratio level, a person who scored zero on the test would have no intelligence at all. And a person who scored 180 would have three times the intelligence as a person who scored 60.

Many physical properties are measured at the ratio level. Height in inches, as measured by a yard stick, is an example of ratio-level measurement. So is weight in pounds, as measured with a bathroom scale.

The second question on our list asked whether you were willing to state that the difference in intelligence between Sally and Joe was the same as the difference in intelligence between Bob and Mary. Had you answered this question affirmatively, you would have been asserting that the IQ test measured intelligence at the **interval level** of measurement. Interval-level measurement assumes that a given interval on the score scale (10 points in this example) represents the same amount of the variable being measured, regardless of where the interval appears on the scale. Unlike ratio-level measurement, interval-level measurement does not imply that zero on the score scale represents an absence of the variable being measured. With interval-level measurement, zero is an arbitrary scale point that has no special meaning.

Temperature, measured in degrees Fahrenheit, is an example of interval-level measurement. The temperature difference between 65 degrees and 70 degrees *is* the same as the temperature difference between 80 degrees and 85 degrees, but zero on the Fahrenheit scale does *not* correspond to absence of temperature. Those of you above the Mason-Dixon line know only too well that zero degrees can seem quite warm after a spell of sub-zero weather. Zero degrees is just an arbitrary reference point.

Had you agreed with our third question, that Bob was most intelligent, Mary came next, Sally was third from the top in intelligence, and Joe was least intelligent, you would have been assuming that the IQ test measured intelligence at the **ordinal level**. Ordinal-level measurement assumes that you can rank-order the objects of measurement, from highest to lowest, on the basis of their scores. To say that Bob is most intelligent does not assume anything about the *size* of difference between his intelligence and Mary's. For instance, it does not assume that the difference between Bob's intelligence and Mary's is the same as the difference between Mary's intelligence and Sally's. Equal differences between ranks do *not* correspond to equal differences in scores. The ordinal-level measurement assumption is weaker than the interval-level assumption. It demands less of the results of measurement.

School grades provide a good example of ordinal-level measurement. An "A" is better than a "B," and a "B" is better than a "C." But the difference between an "A" and a "B" is not necessarily the same as the difference between a "B" and a "C." (When we calculate "grade point averages" we *assume* grades are measured at the interval level, but that's often false; a teacher will often say that a D is (that is, refers to work that is) only a little worse than a C, but an F is much worse than a D.)

If you disagreed with our first three questions and only agreed with the last, you were assuming that the IQ test measured intelligence only at the **nominal level.** You would be stating, in effect, that you would be willing to place the four persons into different intelligence categories, but would be unwilling to state who was more intelligent and who was less intelligent. You can think of nominal-level measurement as the weakest assumption you could make. You would be treating the IQ-scores just like numbers on football jerseys; they serve as surrogate names for the players, but nothing more. They don't represent differences in playing ability or any other variable that is pertinent to the game.

Were we to arbitrarily assign numbers to categories of eye color: 1=blue eyes, 2=brown eyes, 3=green eyes, etc., we would have nominal-level measurement. The numbers wouldn't correspond to any property of eye color, except to say that people who had different numbers were in different eye color categories. For some people, this wouldn't even count as measurement; it would just be description. But, as we'll see in a moment, it's useful to consider it as a limiting case.

We have already suggested that various statistics incorporate particular assumptions about level of measurement. We'll now discuss those assumptions for measures of central tendency. Since the *mode* is merely the score value or category that has the highest frequency associated with it, this measure of central tendency only makes use of the nominal property of data. It can therefore be used with data that represent an underlying variable at any level of measurement—nominal, ordinal, interval, or ratio. Our eye color example is appropriate here. Suppose that you had recorded the eye colors of 20 people, using the scale we described earlier: 1=blue eyes, 2=brown eyes, and 3=green eyes. When you computed the mode of the distribution, you found it to equal 2. You would then know that there were more brown-eyed people in the group than people with any other eye color. In contrast, a mean eye color of 2.57 would be meaningless, since it would not correspond to any eye color.

The *median* assumes that a set of data represents an underlying variable at least at the ordinal level of measurement. The median can therefore be used with data that measure at the ordinal level, the interval level, or the ratio level. If used with nominal-level data, the median will produce a meaningless value. Knowing that the median grade awarded in a class was a "B" is interpretable, since half the students in the class would then have earned a grade of B or below, and half would have earned a grade above a B. However, knowing that the median eye color was "brown" tells you nothing, since eye color is a nominal-level variable, and not an ordinal one.

The *mean* should only be applied to data that measure an underlying variable at the interval level or the ratio level. If the mean is applied to ordinal-level data or to nominal-level data, the resulting value will be meaningless!

When you read research and evaluation reports that include measures of central tendency, we suggest that you ask yourself whether the author's choice of a statistic is consistent with your belief about the level of measurement of the data being described. If an author reports that a group's mean attitude toward their employer is 24.5, ask yourself whether you really believe that the attitude scale measures at the interval or ratio level. Do you believe that the interval between 15 and 20 represents the same difference in attitude as the interval between 25 and 30? If your answer is no, you should question the use of the mean. And the same applies to grade-point averages.

Shape of the Score Distribution. If a score distribution is symmetric and has only one peak, as in Figures 11, 12, and 13 of Chapter 1, the mean, the median, and the mode will have exactly the same value. In this case, other factors should influence an author's choice of a measure of central tendency.

If a score distribution is bi-modal and symmetric, as in Figure 14 of Chapter 1, the mean and the median will have the same value, and the distribution will have two modes. Each mode will correspond to a peak in the distribution. In this case, the modes will not be good indicators of the center of the distribution, but they will describe other important characteristics of its shape. Since the median and the mean will be equal, the choice of either one as a measure of central tendency should be made on the basis of factors other than the shape of the distribution.

If a distribution has only one peak and is not symmetric, the mean, median and mode will all be different. If the score distribution is positively skewed, the mode will be smallest, the mean will be largest, and the median will be between the mode and the mean. This is illustrated in Figure 1.

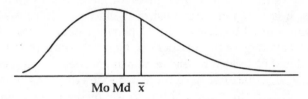

Mo Md $\bar{x}$

Figure 1. Relationship of Central Tendency Measures in a Positively Skewed Distribution of Scores.

If a score distribution is negatively skewed, like the one shown in Figure 2, the mean will be smallest, the mode will be largest, and the median will be in between the two.

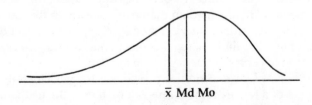

$\bar{x}$ Md Mo

Figure 2. Relationship of Central Tendency Measures in a Negatively Skewed Distribution of Scores.

When a distribution is highly skewed, the mean will not do a good job of representing the center of the distribution; either the median or the mode should be used instead. To see how the mean can give a false indication of the center of a skewed distribution, consider the following example. Suppose that you had a distribution of annual salaries for eleven people. Ten were first-year public school teachers, and one was the late J. Paul Getty, the oil magnate. Suppose that all ten school teachers earned around $14,000 per year, and that Getty earned around $1,000,000 per year. This distribution of annual incomes would be very positively skewed. If we assume that the average income of the ten teachers was $14,000, the mean income for the 11 people would be $14,000 times 10 plus $1,000,000, divided by the sample size of 11. The mean would then be $1,140,000/11= $103,636.36. Now this figure vastly overestimates the average income of the teachers, and greatly underestimates Getty's income. It is nowhere near most of the incomes. In contrast, the median annual income of the 11 people would be close to the $14,000 average of the 10 teachers. Getty's income of $1,000,000 would just be treated by the median as a value that fell above the middle. It would not have the inordinate effect on the median that it did on the mean. Although the median would virtually ignore Getty's very high income, it would do a good job of indicating the center of the distribution of most of the scores.

A similar example of the effect of a skewed distribution on the mean was developed by Alexander Law (1982). Suppose that five people in an office made a contribution to the United Way Fund. Their contributions were $10, $1, $0.50, $0.25 and $0.25. Their mean contribution would be $12.00/5=$2.40, whereas the median of their contributions would be $0.50. Except for that of the one generous contributor, most of the contributions are far better "represented" by the median than by the mean. In this case, the mode of the contributions would be $0.25, which is closer to the center of the distribution than the mean, but wouldn't be as good a measure of central tendency as the median.

Another of Law's examples provides a further illustration of the effect of a few extreme scores on the mean, the median, and the mode. In Table 1, distributions of achievement test scores are shown for two groups, each consisting of 25 students. In the first group (Case 1), the students' scores ranged from a low of 60 to a high of 98. In the second group (Case 2), all students except the three who scored lowest earned the same scores as corresponding students in the first group. In Case 1, the student numbered 23 scored 61; in Case 2, this student scored 21. In Case 1, students numbered 24 and 25 scored 60; in Case 2, these students scored 20. Otherwise, all corresponding scores were identical in the two groups.

Table 1. Achievement test scores for two cases
Results of Achievement Test

			N=25		
Case 1			Case 2		
Student No.	Score		Student No.	Score	
1	98		1	98	
2	98		2	98	
3	97		3	97	
4	81		4	81	
5	80		5	80	
6	79		6	79	
7	78		7	78	
8	77		8	77	
9	75		9	75	
10	75		10	75	
11	73		11	73	
12	72	Range=98-60	12	72	Range=98-20
13	71	Median=71	13	71	Median=71
14	71	Mean=73.6	14	71	Mean=68.8
15	70		15	70	
16	68		16	68	
17	68	Mode=68	17	68	Mode=68
18	68		18	68	
19	67		19	67	
20	65		20	65	
21	65		21	65	
22	63		22	63	
23	61		23	21	
24	60		24	20	
25	60		25	20	

Notice that the median is exactly the same for both groups of students (the median score is 71). The mode is also the same for both groups (the mode is 68). However, changing only the lowest three scores in a group of 25 has caused the mean to drop from

73.6 for Case 1 to 68.8 for Case 2. This drop of almost five points shows clearly the effect that a few extreme scores can have on the mean. As Law noted, in Case 2 sixty percent of the students (15 out of 25) scored 70 or better. Yet the mean of this score distribution is only 68.8. With a distribution that is highly skewed, as in Case 2, the median does a far better job of representing the center of the distribution than does the mean.

Stability. Very often in research and evaluation work, an author will try to generalize his/her results beyond the sample of individuals who have been observed or measured. For example, when two mathematics curricula are being compared in an experiment, an author may conclude that Curriculum A is better than Curriculum B, not only for the samples of sixth-graders used in his/her experiment, but for sixth-graders in general. This is a form of inference that will be discussed in great detail in latter chapters of this book. However, it is relevant to the choice of a measure of central tendency, even when formal statistical inference is not used.

If a researcher really wanted to know whether Curriculum A was better than Curriculum B for "all" sixth-graders in a school system, he or she would, ideally, test the two curricula with every sixth-grader in the system. Usually, such large-scale experiments are impractical and are far too costly. What is done instead, is to select representative samples from a list of all sixth-graders in the system, and then to use each curriculum with one of the samples. Measures of central tendency (usually means) on some outcome variable (typically a test score) are then compared, to decide which curriculum is better.

In this kind of experiment, average scores on the outcome variable for the *entire* population of sixth-graders in the school system are really the center of interest. Corresponding averages for the *samples* of sixth-graders are merely estimates of the population values. When two curricula are compared on the basis of sample averages, there is always the possibility that one average is higher than the other, not because one curriculum is better than the other, but because a particularly bright sample of students happened to be selected for that curriculum. Had other samples been chosen, perhaps Curriculum B would have been found to be better. To guard against drawing false conclusions, it is necessary to use representative samples, of course, but *also* measures of central tendency that do not fluctuate wildly from one sample to the next.

It is well known that some measures of central tendency are more *stable* than others, in the sense that they fluctuate less, from one sample to another. The mean is the most stable of the measures of central tendency we have discussed. The median is less stable than the mean, but is far more stable than the mode.

You may be surprised by this, in the light of our last example, where changing just three scores moved the mean substantially but did not affect the median. However, that was an artificial example, designed by us to make a key point. Taking real samples from a real population, we would find the median varying *more* than the mean.

Of the three measures of central tendency, the mode is, by far, the least stable. If independent samples of sixth-graders were selected from the population of all sixth-graders in a school system, the modes of their score distributions on an achievement test would fluctuate a great deal, from one sample to the next. The median would fluctuate somewhat, but not nearly as much as the mode. The mean would fluctuate least, just due to differences in the abilities of the sampled students. The mean is therefore the most stable of these measures of central tendency.

Summary

In this chapter we have tried to give you a sense of the meaning of three statistics that refer to "the center" of a score distribution. These measures of central tendency are defined differently, and will, therefore, often differ from each other in value.

Each of the statistics is a good measure of central tendency in some situations, and a bad measure in others. As a reader of research and evaluation reports, it is important that you know when a particular measure of central tendency has been used appropriately, and when that statistic can give a misleading indication of the center of a score distribution.

If a score distribution is reasonably symmetric, and measurement is at the interval level or the ratio level, the mean is the best measure of central tendency. It is the most stable of the measures, and will therefore do the best job of estimating a population average. If a distribution is highly skewed, and measurement is at the ordinal level, the interval level or the ratio level, the median is the best measure of central tendency. It will do a better job of indicating the center of the distribution than will the mean. The median is better than the mode in this situation because it is a more stable statistic. If measurement is at the nominal level, and neither ordinal-level, interval-level, nor ratio-level measurement can be assumed, the mode is the only measure of central tendency that will make sense. Neither the mean nor the median will be useful in this situation.

In practice, it is often worthwhile to compute more than one measure of central tendency when several measures meet the assumptions of the data. If a distribution has more than one mode, the mean or the median will do a better job of describing the middle, but reporting the modes will let a reader know which scores have the highest frequencies. If a distribution is skewed, reporting both the mean and the median will often convey the degree of skewness.

Problems — Chapter 2

2.1 Suppose that you wanted to project your future annual administrative budget, adjusted for inflation, on the basis of your administrative expenditures during the past ten years. For nine of the past ten years, your budget fell between $90,000 and $110,000, but during one year you spent $210,000, including the cost of a major revision of your accounting systems by an outside consulting firm.

To project future administrative budget needs, would you likely be more accurate if you used:

a. the average of the annual expenditures over the past ten years?

b. the mode of the annual expenditures over the past ten years?

c. the smallest of the annual expenditures over the past ten years?

d. the median of the annual expenditures over the past ten years?

Explain the logic underlying your answer to this problem.

2.2 Recently, forty thousand high school juniors in a midwestern state were given a reading competency test as a high school graduation requirement. When the 120-item test was scored, it was found that the mean equalled 95.6, the median score was 98.2, and the mode of the distribution of scores was 101.4.

From these statistics, what would you conclude about the shape of the distribution of scores on the test?
   a. The distribution was negatively skewed.
   b. The distribution was symmetric.
   c. The distribution was positively skewed.
   d. The shape of the distribution cannot be inferred from the information given.

2.3 The distribution of payoffs at a race track is positively skewed, with a mode of $2.00, a median of $2.10, and a mean of $3.75. Given these statistics, which of the following are true statements? (Circle the letter of *each* statement that is true.)
   a. More than half the betters get less than a $3.75 payoff.
   b. Half the betters get less than a $2.10 payoff.
   c. Half the betters get less than a $2.00 payoff.
   d. The most frequent payoff is $3.75.

2.4 Suppose that a personnel consulting firm has administered a scale to measure the employee morale of all employees in each of the five departments in your administrative offices. They report that the average score of employees in Department A was 10, and the average score of employees in Department B was 21. In their conclusions, they warn you that "The morale of employees in Department B was more than twice as high as that of employees in Department A."
   a. Is an average the appropriate statistic to use in summarizing the central tendency of the employee morale data?
   b. Should you accept the conclusion that the morale of employees in Department B was more than twice as high as that of employees in Department A? Why might you challenge this conclusion?
   c. Would you accept the conclusion that the average morale of employees in Department B was higher than that of employees in Department A? What measurement assumption would you have to make to accept this conclusion?

3.

Measures Of Variability

How Spread Out Are the Scores?

Although statistics that indicate the center of a score distribution provide useful and important information about a group of scores, they certainly don't give a complete picture of the distribution. Two score distributions that have identical means, medians, and modes, such as those shown in Figure 1, can still differ from each other in important ways.

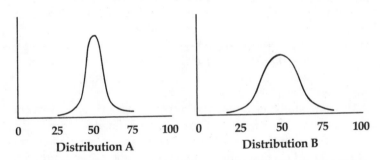

| 0 | 25 | 50 | 75 | 100 | 0 | 25 | 50 | 75 | 100 |

Distribution A   Distribution B

Figure 1. Two Score Distributions That Are Equal in Central Tendency, but Differ in Variability

The distributions shown in Figure 1 are identical in their central tendencies. In addition, both are symmetric and both are unimodal. They differ only in the spread of their scores, the extent to which the scores differ or vary.

The statistics described in this chapter indicate the spread of scores in a distribution. They are called **measures of variability.** Each of the statistics we'll describe would have a larger value for Distribution B in Figure 1 than for Distribution A. As was true of the measures of central tendency discussed in Chapter 2, the most frequently used measures of variability have different properties, and provide the best indication of the spread of scores in a distribution in different situations. In this chapter we first define several measures of variability, we then describe their properties, and finally discuss the appropriateness of the measures in various situations. We hope you'll be a better informed reader of research and evaluation reports when you finish this chapter. Remember that high variability reflects large individual differences, so you *must* know about it before drawing any conclusions about individuals in a group from information about the group's average performance, whether mean, median or mode.

Measures of Variability

The simplest measure of the spread of a group of scores is called the **range.** The range is simply the difference between the largest score in a group and the smallest score in the group. For example, in Table 1 of Chapter 2, Student Number 1 of Case 1 scored 98 on an achievement test. Student 25, the lowest-scoring student in this group, scored 60. For Case 1, then, the range is equal to 98−60=38. For Case 2 in Table 1, Student Number 1 scored 98 and Student Number 25 scored 20. The range of scores for Case 2 is thus 98−20=78.

In some statistics texts, and in some research and evaluation reports, you may find a range that is defined as (the highest score−the lowest score)+1. This slight variation in the definition of range is of no practical consequence. It merely accommodates the real limits of the highest and lowest scores in a distribution. For example, in Case 1 of Table 1 in Chapter 2, the highest score, 98, has real limits of 97.5 and 98.5. The lowest score has real limits of 59.5 and 60.5. If the range were defined as the difference between the upper real limit of 98 and the lower real limit of 60, it would equal 98.5−59.5=39. This is the same as (98−60)+1=39. This range is called the **inclusive range,** and the range that is defined as the difference between the highest and lowest scores in a distribution is called the **exclusive range.** As we noted, these two definitions of the range are important only to statistical purists. For practical purposes either definition will suffice. We have mentioned both only because you might come across them in your study of the research in your field.

In Chapter 1 we introduced the concept of percentiles: statistics that divide a score distribution into two parts, so that a specified percentage of the distribution falls below the chosen statistic. Recall that we defined the 25th percentile, or first quartile, as the point on a score scale that divided a distribution so that 25 percent of the scores were less than or equal to that score value. We also defined the 75th percentile, or third quartile, as the point on the score scale that had 75 percent of the scores in a distribution less than or equal to it.

Two useful measures of variability can be defined in terms of the 25th percentile and the 75th percentile. One such measure is the **interquartile range,** which in some modern statistics texts is called the **hinge spread.** The interquartile range is equal to the difference between the 75th percentile and the 25th percentile. It indicates the spread of the middle half of a score distribution. Another, more widely used measure of variability is the **semi-interquartile range.** The semi-interquartile range is equal to half the difference between the 75th percentile and the 25th percentile. It is found merely by subtracting the 25th percentile from the 75th percentile, and dividing the difference by two. In most statistics texts and research reports, the letter "Q" is used to denote the semi-interquartile range. These measures are less sensitive than the range to the behavior of one or two exceptionally high or low scores, so, like the median, they give a better picture of the distribution for many purposes.

At this point in your statistical travels, if you read that a researcher had administered a 50-item test to a group of fourth-graders, and their median score was 35, you'd probably have some sense of where their distribution of scores was centered. However, if you read that the semi-interquartile range of their scores was 5, you might not know whether their score distribution was widely spread or very narrow. The following rule might help. If the score distribution was reasonably symmetric, half the scores in the distribution would fall between a score value equal to the median plus the semi-interquartile range (Mdn. +Q) and a score value equal to the median minus the semi-interquartile range (Mdn.−Q). In this example, the upper value (Mdn.+Q) =35+5=40; and the lower value (Mdn.−Q)=35−5=30. So if their score distribution was nearly symmetric, you would know that half the students scored between 30 and 40 on the test. So that's a pretty narrow distribution.

What if the distribution wasn't symmetric? In that case, *more* than half the scores would fall between (Mdn.+Q) and (Mdn.−Q). In a highly skewed distribution, as many as 70 percent of the scores might fall between these two values. So the (Mdn.+Q) and the Mdn.−Q) define values that will include *at least half* of a score distribution.

The most widely used measure of variability is the **standard deviation.** This statistic is intimately related to the mean, and is almost always used as a measure of variability when the mean is used as a measure of central tendency. In fact, the standard deviation is a measure of the spread of the scores *around the mean*. The more the scores differ from the mean, the larger the standard deviation will be. If all of the scores in a distribution were equal (e.g., everyone scored 25 on a test), each score would equal the mean and the standard deviation would be zero. If everyone scored between 24 and 26, the standard deviation would be greater than zero, but it would be a small number, probably less than 1. If half the students scored zero on a 50-item test, and the other half earned perfect scores of 50, the standard deviation would be a large number, because every score is a long way from the mean (which is 25).

The standard deviation is related in a very simple way to another measure of variability that is used quite often in inferential statistics, but very little in descriptive statistics. This statistic is called the **variance.** The standard deviation is equal to the square root of the variance. For example, if a distribution of scores had a variance of 16, the standard deviation would equal 4. So the standard deviation can always be found from the variance, and the variance can always be found from the standard deviation. To find a variance from a standard deviation, merely square the standard deviation (multiply it by itself). For example, if a distribution of scores has a standard deviation of 6, its variance will equal 6 times 6=36.

In many research and evaluation reports, you will find the standard deviation denoted by the letter "s," or the Greek version of the letter s, which is called sigma. The variance is often denoted by "s-squared"; that is, an "s" with a 2 printed half a line above it: s^2.

Interpretation of a standard deviation presents the same sort of dilemma as interpretation of a semi-interquartile range. What is a big one and what is a small one? If the standard deviation equals 10, is the distribution narrow or really spread out? Just as the semi-interquartile range was interpreted in conjunction with the median, the standard deviation is interpreted in conjunction with the mean. Here's an example. Suppose that, on a 100-item test, the mean score was 60 and the standard deviation was 10. We can interpret these results in several ways. First, for any score distribution, regardless of its shape, *at least* 75 percent of the scores will fall in an interval that ranges from two standard deviations below the mean to two standard deviations above the mean: $(X-2s)$ to $(X+2s)$. In this example, we therefore know that *at least* 75 percent of the students scored between $[60-2(10)]=60-20=40$ and $[60+2(10)]=60+20=80$. Another rule that applies to virtually all score distributions, regardless of their shape, is the following: *At least* 89 percent of the scores in the distribution will fall between score values that are three standard deviations below the mean and three standard deviations above the mean: $(X-3s)$ and $(X+3s)$. In this example, then, *at least* 89 percent of the students will score between $[60-3(10)]=30$ and $[60+3(10)]=90$. We have underlined the words "at least" in stating and applying these rules because the percentages given are *very* conservative. In *most* score distributions, the percentage of scores that fall within two standard deviations of the mean (no more than two standard deviations below the mean and no more than two standard deviations above the mean) is far greater than 75 percent. Likewise, you will likely find that *far* more than 89 percent of the scores in a distribution will fall within three standard deviations of the mean.

This leads us to the second set of rules that apply only when a score distribution looks like the bell-shaped normal distribution we discussed in Chapter 1. If a score distribution is shaped like a normal distribution, the following rules apply: First, a little more than two-thirds of the scores in the distribution (68 percent, to be more precise) will fall within an interval that is one standard deviation below the mean to one standard deviation above the mean. Second, slightly more than 95 percent of the scores will fall within an interval that is two standard deviations below the mean to two standard deviations above the mean. Finally, 99.7 percent of the scores in the distribution (almost all of them) will fall within an interval that ranges from three standard deviations below the mean to three standard deviations above the mean. These rules are summarized in the following table, and illustrated in Figure 2.

Some Characteristics of a Normal Distribution

Interval	Percent of Scores Included
(X−s) to (X+s)	68
(X−2s) to (X+2s)	95
(X−3s) to (X+3s)	99.7

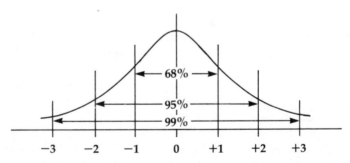

Figure 2. The Bell-Shaped or Normal Distribution Curve.

Let's go back to our example of a 100-item test, administered to a group that had a mean score of 60 and a standard deviation of 10. If we assume that the distribution of scores is *normal* in shape, we can draw the following conclusions: First, 68 percent of the students have test scores that are between (60−10)=50 and (60+10)=70. Second, 95 percent of the students have test scores that are between [60−2(10)]=(60−20)=40 and [60+2(10)]=80. Finally, 99.7 percent of the students have test scores that are between [60−3(10)]=(60−30)=30 and [60+3(10)]=90. Let's summarize these data in a table like the one above:

Interval	Percent of Scores Included
50 to 70	68
40 to 80	95
30 to 90	99.7

From these data, you can see that knowing the mean and standard deviation of a distribution of scores tells you a good deal about the entire distribution. If you can assume that the distribution of scores is bell-shaped (and many distributions are), it is easy to define score intervals that will contain various percentages of the distribution.

Properties of the Measures

As was true for measures of central tendency, the measures of variability we have defined have different properties and are appropriate in different situations. The three considerations we applied to measures of central tendency in Chapter 2 also apply to measures of variability when deciding whether an author has used the right statistic. In that chapter we listed (1) level of measurement, (2) shape of the score distribution, and (3) stability, as criteria to use in evaluating the appropriateness of a measure.

Level of Measurement. As was the case with the measures of central tendency discussed in Chapter 2, various measures of variability make use of different measurement properties of the data to which they are applied. Like the mean, the standard deviation should be applied only to data that measure an underlying variable at the interval level or the ratio level of measurement. The same is true of the range and the variance. Because it is based on percentiles, the semi-interquartile range will provide an interpretable measure of variability if it is used with data that measure at the ordinal level, the interval level, or the ratio level. There is no widely-used measure of variability that is appropriate for nominal-level data.

Shape of the Score Distribution. If a score distribution is reasonably symmetric and unimodal, and measurement is at the interval level or the ratio level, the standard deviation is the most appropriate measure of variability. If a distribution is highly skewed, scores that are in the tail of the distribution (the part that has low frequencies associated with it) will be far from the mean, and will cause the standard deviation to become quite large. This large standard deviation will not give an accurate indication of the spread of scores in the distribution. In this case, the semi-interquartile range will be more appropriate than the standard deviation as a measure of variability, *provided* the scores in the distribution measure an underlying variable at least at the ordinal level of measurement.

Another situation in which the semi-interquartile range is more appropriate than the standard deviation is when a distribution is reasonably symmetric, with most of the scores concentrated around a central peak but with smaller concentrations of scores both well below, and well above, the central peak. For example, if most of the scores in a distribution fell between 40 and 60, but there were small concentrations of scores at 25 and 75, the standard deviation would be greatly affected by the few extreme scores. The standard deviation would become far larger than necessary to represent the variability of the scores that were concentrated between 40 and 60.

Stability. The standard deviation is the most stable of the measures of variability we have discussed. If independent samples were selected from a population and all of the measures of variability we have discussed were to be computed for each sample, the standard deviation would fluctuate least from sample to sample. This is very important when a sample standard deviation is used to estimate the standard deviation for a population. The semi-interquartile range is not as stable as the standard deviation. Its value would change more, from one sample to the next, than would the standard deviation. However, other advantages of the semi-interquartile range outweigh its relative instability in many situations.

The range is the least stable of the measures of variability we have defined. In fact, the range is so unstable that it should never be used as the sole measure of variability. Since the range is computed from only two scores, the highest and lowest, it is not surprising that it will fluctuate wildly from one sample to the next. In a sample of 1000 scores, the range ignores 998 and makes use of only the two extremes.

Summary

We have described five statistics that indicate the variability of a distribution of scores: the range, the interquartile range, the semi-interquartile range, the variance, and the standard deviation. These five statistics have different properties and are appropriate for different situations.

Because it is the most stable of the measures of variability, the standard deviation is the preferred statistic in those situations where it will provide an accurate indication of the spread of a distribution. The standard deviation should be used only when measurement is at the interval level or the ratio level. It is most appropriate when a score distribution is reasonably symmetric and does not have concentrations of scores that are extremely large or extremely small.

The semi-interquartile range is the preferred measure of variability in a variety of circumstances. If the data measure an underlying variable at the ordinal level, but not at the interval or ratio levels, the semi-interquartile range is best. If a score distribution is highly skewed, or if it is symmetric but has small concentrations of extreme scores, the semi-interquartile range will give the most accurate indication of variability.

The range should only be used as the sole indicator of the variability of a score distribution for very informal purposes. Although it has the advantages of simplicity and computational ease, the range is a very unstable statistic. Because it ignores all but the two extreme scores, the range can be a very misleading statistic.

As was true in our discussion of measures of central tendency, we do not advocate concentrating on one measure of variability and ignoring others. Very often, an author can do a better job of describing his/her score distribution by reporting several measures of variability, such as the range and the standard deviation or the range and the semi-interquartile range.

As you read and review research and evaluation reports, be on guard for the use of misleading measures of variability. Don't draw conclusions about the variability of a score distribution just by considering the range. If you're convinced that the variable under discussion can only be, or has only been measured at the ordinal level, but not at the interval level or the ratio level, don't accept the standard deviation as a measure of variability. Look for a semi-interquartile range instead. Examine frequency polygons or frequency distribution tables to see whether reported measures of variability are reasonable, in light of the shapes of distributions. By thinking about the general appropriateness and plausibility of statistics as well as their numerical values, you will be better informed and less often misled.

Problems—Chapter 3

3.1 Look at the two frequency polygons shown below, for distributions of heights (in inches) of twelve-year-old girls and twelve-year-old boys. Assuming the distributions shown in these figures are correct, which of the following statements is true?

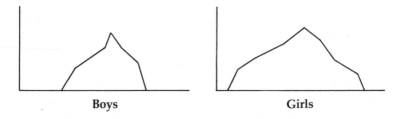

Boys   Girls

    a. The range of heights is larger for twelve-year-old boys than for twelve-year-old girls.

    b. The standard deviation of heights is larger for twelve-year-old girls than for twelve-year-old boys.

    c. The distributions of heights of twelve-year-old girls and boys are about equal in variability.

3.2 Suppose that a report on the effectiveness of an instructional program in mathematics provided the following statistics: When the program was used in 100 different schools, the median of the distribution of school average scores on a post-test of achievement was 28. The semi-interquartile range of the distribution of school average scores was 4. The distribution of school average scores was symmetric.

Between what two school average scores would the middle 50 percent of the distribution of school averages fall?

_____ and _____

3.3 Suppose that teachers in a large metropolitan school system are given contracts of different length, depending on such factors as years of experience, degrees held, rating at last evaluation, and field of specialty. Upon reading a report on personnel practices in that school system, you find that the distribution of lengths of teacher contracts is normal in shape, with a mean of 3 years and a standard deviation of 6 months.

Given these data, which of the following conclusions would likely be true? (Circle the letter of *each* true statement.)

a. Hardly any teachers would have contracts longer than five years.
b. Hardly any teachers would have contracts shorter than one year.
c. The majority of teachers would have contracts with lengths between two years and four years.
d. About half the teachers would have contracts of four years or shorter.

3.4 In a report on the entering freshman class at a local college, the Director of Admissions proudly wrote that the mean rank-in-high-school-class of students admitted in 1982 was 10.6, and the standard deviation of rank-in-high-school-class was 4.2.

a. Do you see anything wrong with using these measures of central tendency and variability in describing the distribution of rank-in-high-school-class of entering freshmen?
b. What level of measurement is assumed when data are reported in terms of ranks?
c. What statistics would be more appropriate for describing the central tendency and variability of a distribution of rank-in-high-school-class?
   Central tendency:
   Variability:

4.

Correlation

The General Idea

A great many questions in applied research and evaluation involve the degree to which two variables "go together." For example, a researcher might ask whether, and to what extent, students who earn high scores on an aptitude test also tend to earn good grades in school. An employer might ask whether, and to what degree, job applicants who earn high scores on a screening test tend to receive high ratings from their supervisors on the job. A psychologist might ask whether, and to what extent, examinees who earn the highest scores on an anxiety scale tend to earn the lowest scores on an examination. All of these questions are concerned with the degree of "co-variation" of two variables or measures. That is, they are concerned with the tendency of the measures to vary together, in the sense that one increases as the other increases (in which case we say that the variables "co-vary positively"), or in the sense that one variable increases as the other decreases (in which case we say that the variables "co-vary negatively"). If they co-vary, one can predict one variable from the other, e.g. college performance from test results.

This chapter is concerned with statistics that indicate the extent to which two variables co-vary. Such statistics are called **correlation coefficients.** In the remainder of this section we will elaborate on the nature of correlation. In the next section, we describe some properties of the most widely used correlation coefficient, called the **Pearson product-moment correlation coefficient.** In a final section, we introduce a number of other correlation coefficients that are used with variables that are measured at the ordinal and nominal levels of measurement, as well as some related statistics. As usual, our discussion will concentrate on the interpretation of results, and ignore the details of computation.

A graph called a **scatter diagram** or **scattergram** provides one of the most useful ways of seeing (or showing) whether, and to what degree, two variables co-vary. In a scatter diagram, one of the variables to be correlated is plotted on the horizontal axis, and the other is plotted on the vertical axis. Each individual's scores or measurements on the two variables are represented by a single point. The point is plotted directly above the individual's score on the horizontal-axis variable, and to the right of the individual's score on the vertical-axis variable.

A typical scatter diagram is shown in Figure 1. In this figure we have plotted the weights (in pounds) and heights (in inches) of 14 hypothetical persons. We have highlighted the point that represents a person who weighs 150 pounds and is 65 inches tall. Notice that the point for this person is plotted just above 150 pounds on the weight scale, and just to the right of 65 inches on the height scale. Try to find the weight and height of the person whose point is closest to the lower left-hand corner of the graph. If you said that this person weighs about 125 pounds and is about 62 inches tall, you'd be right!

If you focus your attention on the entire collection of points in the scatter diagram, you will notice that the points tend to "scatter" upward and to the right. This means that people who weigh more tend to be taller. It also means that taller people tend to weigh more. This relationship can be summarized by saying that for this group of people, weight and height appear to *co-vary positively*. This summary statement does *not* mean that being tall "causes" increased weight, nor that being heavy "causes" increased height. It does mean that heavier people, on average, tend to be taller than lighter people. The "on average" in the last sentence is important. Even though weight and height *tend* to increase together for this group of people, there are several exceptions to the rule. One of our hypothetical people weighs about 140 and is 72 inches tall. Another weighs 165, but is only 70 inches tall. Scatter diagrams in which every person who has a higher score on one variable also has a higher score on the other variable are extremely rare. Exceptions to general trends are found in

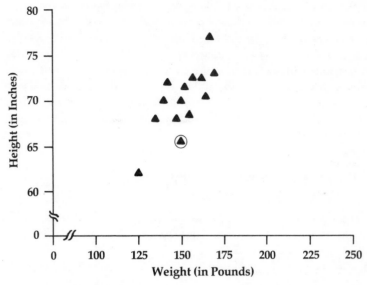

Figure 1. Scatter Diagram Showing Weights and Heights for 14 Persons

virtually all scatter diagrams that represent real data.

We said that although height and weight tend to co-vary positively in the scatter diagram shown in Figure 1, you should not therefore conclude that increased height causes increased weight, nor that increased weight causes increased height. This point cannot be overemphasized. To say that one variable causes another is far stronger than saying that they co-vary, or equivalently, that they "are correlated." It's *usually* true that if two variables do not co-vary, one cannot cause the other, but the fact two variables *do* co-vary does not often indicate a causal connection. Except for the clearly causal relationship between mosquitos and mosquito bites, it is safe to say that nothing is caused solely by a single factor in the real world of research and evaluation. Often, two variables are correlated because both are caused by other same variables. For example, were you to plot a scatter diagram with per-capita alcoholic consumption in the United States on the X-axis, and average teacher salary in the United States on the Y-axis, you would find a very strong positive relationship between these two variables over the years 1920 to

1980. In fact, an increase in alcoholic consumption would almost always be accompanied by a corresponding increase in average teacher salary. But it would be wrong to conclude from this very strong relationship that teachers spent all of their salary increases on demon rum, and just as wrong to conclude that an alcoholically lubricated society was led by its inebriation to raise teacher salaries. More likely, the increase in these two variables over the years was caused by an increase in gross national product, coupled with inflation.

The Pearson Product-Moment Correlation Coefficient

The Pearson product-moment correlation coefficient provides a numerical summary of the kinds of relationships illustrated in scatter diagrams. When this statistic is appropriate, it indicates the degree to which two variables tend to co-vary. More specifically, the Pearson correlation coefficient, "r" (authors rarely use its full name), provides an indication of the tendency of points in a scatter diagram to fall right along a straight line.

If the points in a scatter diagram fall *right on* a straight line that is angled upward, from lower left to upper right, the associated Pearson correlation coefficient is equal to +1.00. If the points in a scatter diagram fall *right on* a straight line that is angled downward, from upper left to lower right, the associated Pearson correlation coefficient is equal to −1.00. (It doesn't matter what the particular positive of negative *angle* of the line is, only that the point falls *on* it.) If the points in a scatter diagram were scattered uniformly throughout a circular pattern, there would be no tendency for one variable to increase or decrease as the other variable increased. The associated Pearson correlation coefficient would then equal zero.

The greater the tendency of points in a scatter diagram to fall along a straight line that is angled upward (a line that has "positive slope"), the closer the Pearson correlation coefficient will be to +1.00. The greater the tendency of points to fall along a straight line that is angled downward (a line that has "negative slope"), the closer the Pearson correlation coefficient will be to −1.00. These relationships are illustrated by the hypothetical scatter diagrams shown in Figure 2.

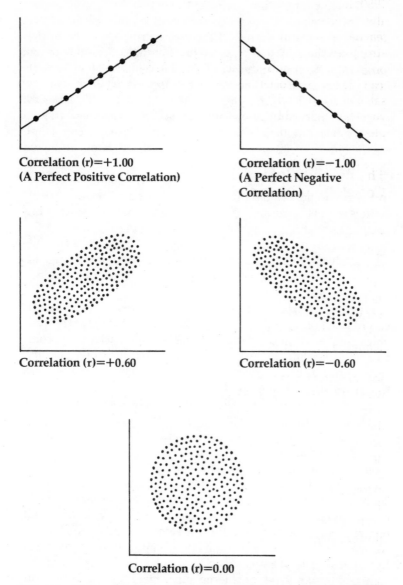

Correlation (r)=+1.00
(A Perfect Positive Correlation)

Correlation (r)=−1.00
(A Perfect Negative
Correlation)

Correlation (r)=+0.60

Correlation (r)=−0.60

Correlation (r)=0.00

Figure 2. Scatter Diagrams Associated with Pearson Product-Moment
Correlation Coefficients of Various Magnitudes

A Pearson correlation coefficient of +1.00 is called a "perfect positive correlation," and a coefficient of −1.00 is called a "perfect negative correlation." These rare ascriptions of perfection stem from the ability of a researcher to predict the value of one variable perfectly, by knowing the value of the other variable (and the values of a couple of constant factors), when the correlation between the variables is either +1.00 or −1.00. But alas, perfection in research and evaluation is as elusive as perfection in other aspects of life. We've never seen a Pearson correlation coefficient of +1.00 or −1.00 outside a statistics book, and we'd be truly amazed if you were to find one.

Prediction of one variable from another is frequently of interest in research and evaluation. And the closer a correlation coefficient is to plus or minus 1.00, the more accurately a reseacher can predict the value of one variable, by knowing the value of the other. The method most often used for prediction is called **regression analysis,** a topic that is discussed at length in Chapter 14. You will find that regression analysis and correlation are intimately related.

Whether a correlation coefficient is considered to be significantly large or small depends on the usual behavior of the variables used to compute it, and on the group for which it is computed. Some statistics texts provide arbitrary rules for evaluating the magnitude of a correlation coefficient, such as: "a correlation that is less than 0.30 is small, a correlation that is between 0.30 and 0.70 is moderate, a correlation that is between 0.70 and 0.90 is large, and a correlation that is greater than 0.90 is very large." Although such rules provide general guidelines to interpretation, they can be misleading in many specific situations. To tell whether a correlation coefficient is large or small, you have to know what is typical. For example, if a standardized reading achievement test were to be administered to a group of ninth-grade students, and another standardized reading achievement test were to be administered to the same students a week later, we would expect the correlation between their two sets of scores to be between 0.90 and 0.95. In this context, a correlation coefficient of 0.80 would be considered quite small, and a correlation of 0.90 would be considered moderate. In contrast, the correlation between high school graduates' scores on any of the widely-used scholastic aptitude tests, and their first-year college grade point averages, is typically around 0.50. In this context, a correlation of 0.70 would be considered very high, and a correlation of 0.40 would be considered quite low. The best way to determine the "typical" size of a correlation coefficient for particular variables in a particular population is to read some of research and evaluation reports in that area, especially those providing an overview of research.

One way researchers often express the strength of the relationship between two variables is by squaring their correlation coefficient. Suppose a group of students was administered a reading achievement test and a verbal IQ test. If the students' reading achievement scores and verbal IQ-test scores had a correlation of 0.80, a researcher might report the squared correlation as 0.80 times 0.80=0.64. This squared correlation coefficient is called a **coefficient of determination.** The coefficient of determination is useful because it gives the proportion of the variance of one variable that is predictable from the other variable. So we might say that 0.64 (or 64 percent) of the variance of the students' reading achievement scores is predictable from their verbal IQ-test scores. If two variables had a correlation of plus or minus 1.00, the corresponding coefficient of determination would equal plus 1.00. This would mean that 100 percent of the variance of one variable would be predictable using the other variable, and vice-versa. This interpretation is consistent with our earlier comments on a perfect correlation of plus or minus 1.00. Conversely, suppose that two variables had a correlation of zero. Then the coefficient of determination would equal zero, suggesting that none of the variance of one variable was linearly predictable from the other variable. This too, is consistent with our earlier comments. If two variables have a correlation of zero, there is no tendency for the points in their scatter diagram to fall along a straight line. Therefore, knowing a person's score on one variable will not help in predicting their score on the other variable. This interpretation of correlation coefficients is discussed in greater detail in the section of Chapter 14 that deals with regression analysis.

When you find correlations in research and evaluation reports, you might wonder *why* particular coefficients are large or small. Unfortunately, the answer is rarely simple. There are many reasons why a correlation coefficient might be large and many reasons why it might be small. Proper interpretation demands thoughtful consideration of many factors, and a good bit of caution. Finding a large correlation coefficient might mean that one of the variables being correlated has a major influence on the other; the relationship between the variables might be causal. Alternatively, a large correlation coefficient might result because the variables being correlated have a number of causes in common. In fact, the variables being correlated might be symptoms of other, underlying variables, that have a more fundamental relationship. For example, the correlation between elementary school students' family income and their verbal intelligence would probably be quite large. But this does not mean that being verbally bright causes students to come from rich families, nor that being from a rich family causes verbal intelligence. Family income is most likely an indicator of an intellectually rich home environment, with many opportunities to practice verbal skills. And a score on a verbal intelligence test is, to some degree, an indicator of the quality of schooling a student has received. The strong relationship between these more fundamental variables might well explain the relatively high correlation between family income and verbal intelligence test scores. Correct interpretation of a large correlation coefficient thus requires knowledge of the real world relationships that may influence the varibles being correlated.

The correlation between two variables might also be large because the data used to calculate it came from several groups that differed on the variables being correlated. For example, the correlation between the weights and reading achievement scores of a group consisting of 20 third-graders, 20 fourth-graders, and 20 fifth-graders would probably be quite high. This is because the fourth-graders would have a higher average reading achievement score and a higher average weight than the third-graders, and the fifth-graders would have a higher average reading achievement score and a higher average weight than the fourth-graders. For students in any one grade, however, the scatter diagram of weights and reading achievement scores would be nearly circular, and the corresponding correlation coefficient would be close to zero. But for students in all three grades, the scatter diagram would be stretched out, and the corresponding correlation coefficient would be moderately large. This is illustrated in Figure 3.

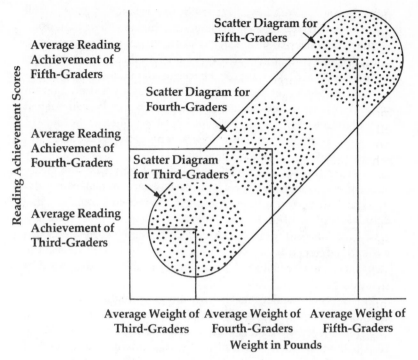

Figure 3. Scatter Diagrams of Weight and Reading Achievement for Students in Grades Three, Four, and Five

When interpreting correlation coefficients in research and evaluation reports, it is therefore important to note the source of the data used to calculate the correlations. The correlation between two variables can be very different, depending on the group that supplied the data, as we have illustrated in Figure 3; it's near zero within one class but high within the school.

Two variables can have a *small* correlation coefficient for many reasons. One possibility is that the underlying variables represented by the available data are related, but not linearly related; that is, although the points in the scatter diagram have no tendency to fall along a straight line, they do have a tendency to fall along a curve. This would indicate a "curvilinear relationship" between the variables, a topic that is discussed in the next section. For example, researchers have found that anxiety and achievement often have a curvilinear relationship. A small amount of anxiety appears to facilitate achievement so that, up to a point, an increase in anxiety is associated with an increase in achievement. But, beyond the point of facilitation, anxiety interferes with achievement, and becomes debilitating. So as anxiety increases further, achievement decreases. A scatter diagram of scores on an anxiety scale versus scores on an achievement test would therefore follow a curve, and not a straight line. Because of this, the Pearson correlation coefficient between anxiety and achievement would probably be low.

Another reason a correlation coefficient might be low is because measurement is poor. If the tests, observation instruments, attitude scales, and other devices used to measure variables like achievement, attitudes, and behavior, do not measure consistently, scores on these devices cannot be highly correlated. You can often detect this problem by examining the **reliabilities** of the devices being correlated. This reliability problem is discussed in Chapter 5.

Yet another measurement problem that might result in low correlation coefficients is called the "floor and ceiling effect." If a test that is very difficult is administered to a group, many of the resulting scores will be very low. Because many examinees will earn nearly-identical low scores, the correlation between scores on that test and scores on any other variable will necessarily be low. To illustrate with an unlikely extreme: If everyone earned the same score on a test, there would be no high scores and no low scores, hence no difference in scores, hence no variance. That test would therefore be useless in predicting differences between scores (variance) on any other variable, and would therefore have a correlation of zero with any other variable. If scores on a test tend to pile up in the low end of the score scale, the test is said to have an "inadequate floor." The test does not measure differences in the ability of low-scorers. A similar problem arises when a very easy test is administered to a group. Many examinees earn the highest possible score, because of the "inadequate ceiling" of the test. Again, the correlation between scores on the test and scores on any other variable will be low. If the difficulty of the test was more appropriate to the ability of the group being tested, the correlation between scores on the test and scores on other variables would undoubtedly be higher. By looking at a scatter diagram, you can often tell whether a variable is exhibiting a floor or ceiling effect.

Other Types of Correlation Coefficients

Although Pearson correlation coefficients are used more frequently than any other kind, they do not give a complete picture of the relationship between two variables, and they are not applicable in all situations. The Pearson correlation coefficient only indicates the degree to which the points in a scatter diagram tend to fall along a *straight* line. This is called the "degree of linear relationship" between two variables. If the points in a scatter diagram tend to fall along a curve, such as an arc (a parabola), the Pearson correlation coefficient might miss their relationship entirely. An alternative correlation coefficient, called **eta,** indicates the degree of relationship between two variables regardless of whether the relationship is linear (i.e., the points in a scatter diagram fall along a straight line), or non-linear (i.e., the points in the scatter diagram fall along a curve). Eta is denoted by the lower-case Greek letter η . A related statistic, **eta-squared,** indicates how well one variable can be predicted from another, either through a linear relationship or a non-linear (curved) relationship. Eta-squared can be compared to the coefficient of determination, to estimate the extent to which two variables have a non-linear relationship. In one of our earlier examples, involving prediction of students' reading achievement scores from their verbal IQ-test scores, the coefficient of determination was found to be 0.64. Suppose that eta-squared for these tests had been found to equal 0.90. That means that 90 percent of the variance of reading achievement scores was predictable from students' verbal IQ-test scores, either linearly or non-linearly; and since 64 percent of the variance of students' reading achievement scores was linearly predictable from their verbal IQ-test scores, 90-64=26 percent of the variance of their reading achievement scores was predictable non-linearly.

We have already suggested that the Pearson correlation coefficient is applicable to data that measure underlying variables at the interval or ratio levels of measurement. Other correlation coefficients have been developed for use with data we can only collect at other levels of measurement. The **Spearman rank-order**

correlation coefficient is widely used with data that measure underlying variables at the ordinal level. The Spearman correlation coefficient is interpreted just like the Pearson coefficient. Although the formulas for the two coefficients are different, the Spearman correlation coefficient should be thought of as just the Pearson correlation coefficient applied to sets of ranks.

If applied to the *same* set of interval-level data, the Spearman correlation coefficient would be somewhat closer to zero than the Pearson correlation coefficient. When it is appropriate, the Pearson correlation coefficient should always be used because it makes use of the interval-level properties of the data. It therefore does a better job of indicating the degree of linear relationship between two interval-level variables. The Spearman coefficient throws away the *size* of the intervals, using only the ranks. But when you only *have* the ranks, you are very happy to have the Spearman coefficient.

In research on testing, students' scores on individual test items are often correlated with their scores on the remainder of the test. Because test items are typically scored zero (for a wrong answer) or one (for a right answer), special correlation coefficients are needed for this application. The **biserial correlation coefficient** and the **point biserial correlation coefficient** meet these needs. Each of these correlation coefficients can be used whenever one of the variables is scored zero or one, and the other variable is measured on an interval scale. The biserial correlation coefficient provides an estimate of what the Pearson correlation coefficient would be, if the researcher had true interval-level measurement for the zero-one variable. This achievement of the biserial correlation coefficient is, to some degree, offset by the fact that it is based on a strong assumption that isn't always met in practice. The point-biserial correlation coefficient takes the zero-one variable as it is, and doesn't incorporate any strong measurement assumptions. When its assumptions hold, the biserial correlation coefficient will usually be larger than the corresponding point-biserial correlation. Some researchers feel that it does a better job than the point-biserial coefficient in indicating the relationship between scores on test items and scores on the remainder of a test. The interpretation of such correlation coefficients is discussed in Chapter 5.

The evaluation of tests provides the context for some additional correlation coefficients. Correlations between examinees' scores on different test items are often useful guides to the improvement of tests. In this application, *both* of the variables to be correlated are scored zero (for a wrong answer) or one (for a right answer). Two correlation coefficients have been developed for variables that are both scored zero or one: the **phi coefficient** and the **tetrachoric correlation coefficient.** The phi coefficient is equivalent to the Pearson correlation coefficient, applied to two variables that are scored zero or one. Because of this scoring, the formula for the Pearson correlation coefficient can be simplified to that of the phi coefficient. The tetrachoric correlation coefficient assumes that the two zero-one items really measure underlying variables that have an interval scale. This coefficient attempts to predict what the correlation would be, if data on these underlying variables were available. Like the biserial correlation coefficient, the tetrachoric correlation coefficient is based on strong measurement assumptions that might not hold in practice. As was true for the Pearson correlation coefficient and all of the other correlation coefficients we have discussed so far, the phi coefficient and the tetrachoric correlation coefficient can assume values between −1.00 and +1.00. The extreme values indicate perfect correlation, and a correlation of zero indicates the absence of a linear relationship.

In many studies, researchers and evaluators attempt to predict one variable (called a criterion variable) from scores on *several* other variables (called predictor variables). For example, students' college grade point averages are often predicted from their high school grade point averages *and* their scores on several aptitude tests (such as the verbal subtest and the quantitative subtest of the Scholastic Aptitude Tests). In these studies, a statistic called the **multiple correlation coefficient** (usually denoted by R) is often used to describe the correlation between actual scores on the criterion variable and predicted scores on the criterion variable. The higher this correlation, the more accurate the prediction. Because of the way it is defined, the multiple correlation coefficient can only assume values between zero and +1.00. If the multiple correlation coefficient were to equal +1.00, each person's predicted score on the criterion variable would equal their actual score on the criterion variable; in that case, prediction would be perfect.

Reports on studies involving multiple correlation sometimes contain a statistic called the **coefficient of multiple determination.** This coefficient is simply the square of the coefficient of multiple correlation, and indicates the proportion of the variance of a criterion variable that can be predicted using all of the predictor variables. For example, a coefficient of multiple correlation equal to 0.80 would have an associated coefficient of multiple determination equal to 0.64. This is sometimes described verbally by saying that the predictor variables "explain" or "predict" 64 percent of the variance of the criterion variable. The larger the coefficient of multiple determination, the better the prediction. The coefficient of multiple determination is usually denoted by R^2. Both the coefficient of multiple correlation and the coefficient of multiple determination are illustrated and discussed in greater depth in Chapter 14.

Summary

Correlation coefficients are statistics that indicate the degree of relationship between two variables. The Pearson correlation coefficient, like most correlation coefficients, indicates only the degree of linear relationship between two variables. In contrast, the eta coefficient indicates the degree of linear and non-linear relationship between two variables.

Correlation coefficients must be interpreted thoughtfully and cautiously. There are many reasons why a correlation coefficient can be large or small. In general, a large correlation coefficient does not mean that two variables are causally related; there are many alternative explanations.

A variety of correlation coefficients has been developed for use with data that measure at the interval, ordinal and two-category nominal levels of measurement. All of these coefficients can take on values between -1.00 and $+1.00$.

Multiple correlation coefficients indicate the degree of linear relationship between a single criterion variable and a number of predictor variables. These coefficients are discussed in Chapter 14.

Problems—Chapter 4

4.1 Suppose that you were to collect data on the ages and weights of 100 persons—both children and adults—who were attending a company picnic. If you were to calculate the correlation coefficient between the ages and weights of these 100 persons, which of the following values is most likely to be the correlation you would find?
   a. A correlation of −0.90.
   b. A correlation of +1.00.
   b. A correlation of +0.02.
   d. A correlation of +0.50.

4.2 Suppose a research report gave the correlation coefficient between the IQ-test scores of students when they were in the second grade and their arithmetic achievement scores when they were in the third grade. The reported value was 0.60. The report also stated that the correlation between the number of days the third-graders were absent from school and their third-grade arithmetic achievement scores was −0.85. Given these data, which of the following are true statements? (Circle the letter of *each* true statement.)
   a. Number of absences during the third grade is a better predictor of third-graders' arithmetic achievement than are their second-grade IQ-test scores.
   b. Number of absences during the third grade is not as good a predictor of third-graders' arithmetic achievement as are their second-grade IQ-test scores.
   c. Second-grade IQ-test performance and number of absences during the third grade are equally good predictors of third-graders' arithmetic achievement.
   d. If a scatter diagram were drawn, using data on third-graders' absences and their arithmetic achievement scores, the scatter of points would go from the upper left corner of the graph to the lower right corner.

4.3 A report on the relationship between expenditures for public education and student achievement in school systems states that "the correlation between annual per-pupil expenditures and average student achievement test performance equals +2.7 for the 500 school systems in the sample." From this result, which of the following can you safely conclude is true?

   a. There was either an error in computing the correlation coefficient or a typographic error in reporting the result.

   b. Higher student achievement attracts higher expenditures for public education.

   c. There is a positive association between per-pupil expenditures for education and student achievement.

   d. Spending more on education causes student achievement to increase.

4.4 Following is a scatter diagram that illustrates the relationship between level of debt and number of employees for a sample of Fortune 500 corporations. Assuming the data shown in the scatter diagram are representative of the entire Fortune 500 list, which of the following correlation coefficients is most likely to summarize the true relationship between debt level and number of employees for those corporations?

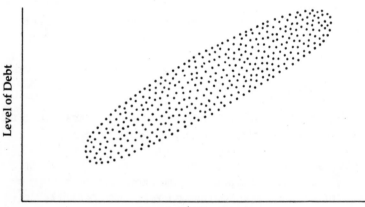

   a. +0.00
   b. +0.60
   c. −0.60
   d. +1.00

5.

Some Fundamentals of Measurement

Measurement as the Basis for Statistics

Researchers in the physical and biological sciences have several distinct advantages, compared to their brethren in the social and behavioral sciences and those who conduct educational research. One major advantage is that physical and biological scientists can often observe the things they want to measure, while social and behavioral scientists often cannot. In educational research, most of the really important and interesting variables are *not* directly observable: many are psychological or social states—the one too private and the other too difficult to observe directly. As you might imagine, not being able to see what you want to measure often creates thorny problems. In the terms of the title of this book, think of how difficult it would be to measure a professional baseball player's hitting ability if it was impossible to observe his performance at the plate. That's how it is most of the time in educational research, and much of the time in social and behavioral research.

Think of the variables most often used by educators to describe the behavior or performance of students. They talk about things like "intelligence," "motivation," "attitude," "aptitude," and "ability." None of these variables can be observed directly. We can stand a student next to a door frame and measure his or her height merely by comparing the distance from the floor to the top of the student's head with the distance shown by a tape measure. But we cannot measure a student's intelligence that way.

Intelligence, and all of the other variables in our list, are examples of what behavioral scientists call **constructs.** They are "constructed variables." We label a person as intelligent because that person exhibits a certain consistent set of observable behaviors. Intelligence is the label applied to the concept underlying the entire set of behaviors. For example, intelligence is suggested by lucid speech and a large vocabulary, but vocabulary alone is not sufficient. Since the turn of the century, a most widely accepted measurement of intelligence is a person's performance on the set of problems that comprise an "IQ-test." It gives us a chance to observe a person's responses to a set of problems that don't have obvious solutions. Collectively, those responses give an indication of the person's status with respect to the construct we call "intelligence."

Motivation is another key psychological construct. We cannot observe a student's motivation directly. However, we can observe that the student always comes to class on time and rarely misses a day of class, always completes assigned work on or ahead of time, always provides a careful and complete response to assigned work, often volunteers to do extra work, and usually volunteers answers to teachers' questions in class. To be sure, variables other than motivation, such as ability and health, affect these behaviors. But, taken together, we would normally describe a student who exhibited all of these behaviors as being "highly motivated." Thus we summarize these and other patterns of consistent behaviors under the label "motivation."

The use of constructs (rather than directly observable variables) in research and evaluation results in a host of measurement issues and, often, problems. First, we have to concern ourselves with the extent of the relationship between certain of the variables we observe (e.g. performance on a vocabulary test) and the constructs we want to measure (e.g. intelligence). This is a **validity** issue. Second, we have to concern ourselves with the consistency of whatever measurement approach we use; e.g., are we measuring motivation in such a way that a student who is determined to be "highly motivated" when one person does the measuring will be similarly assessed by someone else using the same approach? This is a **reliability** issue. Finally, we must concern ourselves with the development of appropriate **units** of measurement. It might be obvious that—in the U.S. today—we should measure students' heights in feet and inches. But what units should we use for measuring motivation? Clearly, feet and inches won't do.

In the remainder of this chapter, we consider the issues of validity, reliability, and units of measurement in greater detail. If you are to be a knowledgeable judge of the results of research and evaluation studies in education and the social and behavioral sciences, it is essential that you learn how to judge the appropriateness and quality of the measurements that are the foundation of such studies. The raw data of educational and behavioral research are only as good as the processes used to produce them. Measurement plays a critical role in all such research. And we have by now developed some very useful statistical tools we can apply to evaluating various approaches to measurement.

Assessing Validity

When we collect data on a person for the purpose of research in education or the social sciences, we usually measure or observe a *sample* of that person's behavior. We might provide the person with a task to be completed or some set of stimulus materials that require a response. A traditional pencil-and-paper test is a good example. Alternatively, we might just observe the person in some naturally-occurring situation, and count the incidences of some behavior or action. In recent years, much educational research has included observations of the incidence of specified students' or teacher's behaviors in classrooms. For example, a researcher might count the number of times, in a 15-minute period, that a teacher asks his or her students a direct question.

If research merely involved reporting the number of test items a person answered correctly, or the number of times a teacher asked his or her students a question, there would be few validity issues. Validity concerns usually arise because researchers interpret these observable behaviors as indicators of a person's status on more general constructs such as "ability" or "interest." By generalizing beyond observed behaviors that occur at a particular time and in a particular setting, researchers and evaluators create a multitude of validity issues. The most fundamental concern is the appropriateness of the interpretations of the results of measurement. Measurement procedures are not *inherently* valid, nor inherently invalid. It is the *interpretations* of measurement results that must be examined for validity.

The nature of measurement validity has been debated for decades, and many issues are still far from being settled. Entire texts have been devoted to the theory of measurement, with discussion of the nature and assessment of measurement validity filling many of their pages. Our discussion here will be relatively brief, and concerned more with practical interpretation than with theoretical underpinnings.

Classical measurement books usually identify four types of measurement validity: **construct validity, concurrent validity, predictive validity,** and **content validity.** Recent contributors to the measurement literature, such as Messick (1980; 1981) have suggested that content validity is really not concerned with measurement validation, and that the other three distinctions are about as confusing as they are helpful. Nevertheless, we think that the four types of validity form a useful framework for critical reading of many research and evaluation reports. We will therefore discuss each one.

Construct validity is at once the most fundamental and the most complex of the four types of validity we have mentioned. (Messick (1980) suggests that it encompasses all of validity, and subsumes the other three types.) Construct validity is concerned with the total relationship between the results of a particular measurement or set of measurements and the underlying construct the researcher or evaluator is attempting to measure. In short, it addresses the question: "Does this measurement procedure really measure the construct it is claimed to measure?" For example, the question: "Does this IQ-test really measure intelligence?" is asking about construct validity. As you might imagine, the question is very difficult to answer, and a conclusive answer is rarely found. All a researcher can do is try to amass a balance of evidence in support of the claim that the IQ-test measures intelligence. Construct validity is never "proved"; it is merely accepted as long as supporting evidence strongly outweighs contrary evidence. A construct validity study is designed to determine whether a measurement procedure yields results that are consistent with theoretical expectations. For example, the theory of intelligence suggests that older children are, on average, more intelligent than younger children. If an IQ-test is truly a measure of intelligence, older children should therefore earn a higher average score on that test than will younger children—and, indeed, they do. Another theoretical expectation is that more-intelligent children, on average, achieve better grades in school than do less-intelligent children. The correlation between children's scores on the IQ-test and their grade-point averages in school should therefore be positive. If the correlation is positive, the construct validity claim—that the IQ-test measures intelligence—is somewhat supported; if the correlation is close to zero or is negative, the claim to construct validity would have suffered a near-fatal blow.

No single piece of evidence is sufficient to establish the construct validity of a measurement procedure. A series of studies that examines every reasonable alternative interpretation of the results of measurement is necessary, before a construct validity claim is adequately supported.

When you read research and evaluation reports that contain discussions involving constructs, it is essential that you question the validity of the authors' claims. If authors state that one group had more-positive attitudes than another, or that students in one academic program are less highly motivated than students in another, you must ask how these constructs were measured and critically question the authors' evidence that their procedures did actually measure attitude and motivation.

At this point, you're probably wondering how you can possibly know whether the construct validity of a measurement procedure has been adequately demonstrated. Unfortunately, there is no simple answer. In any study, you can quickly note whether the author has even addressed construct validity issues—that is, whether there's any explicit discussion of the problem. When a researcher has used a commercially-available measurement instrument, such as a test or an attitude scale, you can check such sources as O.K. Buros' *Mental Measurements Yearbooks*, the *Journal of Educational Measurement*, or *Educational and Psychological Measurement*, for professional reviews that often include discussion of measurement validation. As in other aspects of research and evaluation, healthy scepticism on the part of the reader is fully warranted.

When a researcher has developed a new measurement procedure, he or she might provide evidence of its validity through a method known as **concurrent validation.** Let's consider an example. The Torrance Test of Creativity is one of the most widely accepted measures of what psychologists call "creativity" or "divergent thinking." Suppose that, in an evaluation of an educational program designed to foster creativity in children, you develop your own measurement procedure for "creativity." As evidence supporting this claim, you administer your measurement procedure *and* the Torrance Test of Creativity to a large sample of students who have participated in the new educational program. You then compute the correlation coefficient between students' scores on your test and their scores on the Torrance Test. You find that the correlation coefficient equals 0.90, suggesting that students who earn relatively high scores on your procedure also earn relatively high scores on the Torrance Test, and that students who earn relatively low scores on your measure also earn relatively low scores on the Torrance Test. Since the Torrance Test is an accepted measure of creativity, you can legitimately reason that your procedure also measures that construct.

The essential features of this example are common to all concurrent validation studies. Such studies involve an accepted measure of some construct, and a new, as yet unvalidated measure. The concurrent validity of the new measure is estimated by administering it, together with the accepted measure, to an appropriate group of subjects. The correlation coefficient between scores on the new measure and scores on the accepted measure is known as a **validity coefficient.** The closer the validity coefficient is to 1.0, the higher the concurrent validity of the new measure. Concurrent validation gets its name from the almost-simultaneous administration of the accepted measure and the new measure. In fact, the time interval between the two administrations need only be short enough that no systematic changes in the examinees' behavior are likely to take place. Concurrent validity is often part of a more general study of the construct validity of a new measurement procedure. It should seem reasonable that a new measure of a given construct should give results that co-vary positively with older, accepted measures of that construct. High concurrent validity coefficients thus provide supporting evidence of construct validity.

The primary objective of many applied studies is the *prediction* of subjects' future performance or behavior. For example, a professional polling organization, such as the Gallup Poll, might survey registered voters several months prior to an election in an attempt to predict the election outcome. Such predictions are effective only to the degree that voters' responses to the survey questions indicate their future voting behavior. In an educational context, most selective colleges require that their applicants for admission complete an aptitude test, such as the *Scholastic Aptitude Tests (SAT)* or the *American College Testing Program Tests (ACT)*. Colleges then use the applicants' test scores to predict their future college grade point averages (GPA), and only admit those applicants who are likely to exceed some minimum GPA.

In studies such as these, the correlation coefficient between the measure used for prediction (such as the score on the voters' survey questionnaire) and the measure used to indicate future performance (such as actual voting behavior at the time of the election) is called a **predictive validity coefficient.** The higher this correlation between the predictor variable (e.g., the voters' responses to the survey questionnaire) and the criterion variable (actual voting behavior), the more valid is the predictor.

Predictive validity is of particular importance in situations involving selection of applicants, placement of applicants into appropriate positions (such as occupational specialties within the military or in industry), and in marketing studies that attempt to predict consumer behavior. As is true for all types of validity, a measurement procedure does not have any *inherent* predictive validity. Its predictive validity might be high when predicting some criterion variables, and low when predicting others. It might also be high for some groups of individuals and low for others. For example, it might be the case that people who say they intend to vote Democratic really do so a month later, whereas, people who say they intend to vote Independent often change their minds. In that case, the predictive validity of a survey questionnaire would be high for Democrats, but low for Independents.

When interpreting a research study that involves predictive validity, make sure that the criterion variable and the group used to establish predictive validity are relevant to your own interests and likely applications. Otherwise, you should regard evidence of predictive validity as being tentative for your own purposes.

Issues of **content validity** arise frequently in educational research and evaluation studies, but also in psychiatry and market or political research. In such studies, researchers or evaluators often want to generalize from an individual's (often a student's) performance on a particular test to that individual's ability to solve test problems in some larger **content domain.** For example, suppose that a test contains 20 items (a problem on a test is called an **item**) in which a student is shown four words. In each item, three of the words are spelled correctly and one is spelled incorrectly. The student must identify the word that is spelled incorrectly. In most uses of such a test, the researcher or evaluator would not be interested in the student's ability to discriminate between the *particular* 60 correctly spelled and the *particular* 20 incorrectly spelled words that make up that test. What is really of interest is the student's ability to discriminate between *all* correctly or incorrectly spelled words that are, in some way, like the words used to make up the test. Researchers and evaluators routinely generalize from students' performances on a few specific test items to their hypothetical performance on all possible similar test items. The content validity issue focusses on exactly these generalizations. To be content valid, the items on the test must constitute a **representative sample** of the domain of items used to generalize the students' performances. For example, if all the words used in the 20 spelling items consist of only two syllables and begin with a vowel, it would clearly be unreasonable to generalize the students' performance to all words in Webster's *New World Dictionary*. The test items would not constitute a representative sample of the words in the dictionary.

Unlike construct validity, concurrent validity, and predictive validity, which are assessed at least substantially on the basis of statistical evidence (such as correlation coefficients), content validity is assessed solely on a judgmental basis. The fundamental issue is whether the content of the test is consistent with the content of the domain of items or tasks for which the generalization is claimed to hold. In pursuing this fundamental question, one might ask whether *each type* of item for which representativeness is claimed is actually represented in the test. One might also ask whether each type of item is present in the test *in proportion to* its presence in the domain of content used for generalization. For example, if a mathematics test is claimed to measure students' abilities to solve computational problems involving the four basic arithmetic operations of addition, subtraction, multiplication, and division, then that test should not consist of 80 percent addition items, 10 percent subtraction items, five percent multiplication items, and five percent division items. The four types of items would normally make up approximately equal portions of the test.

A good way to begin analyzing the content validity of a commercially-available test is to obtain the test publisher's technical manual for the test. Most reputable test publishers provide a **test blueprint** or a **table of specifications** that shows how many items in the test were designed to measure each of the types of behavior the test is intended to assess. Comparison of the test blueprint with a desired domain of content is a useful beginning in assessing a test's content validity for a particular application.

A recent court case, *Debra P. v. Turlington* in the State of Florida, has generated still another type of validity in educational testing applications. In an early court ruling, since reversed, the state of Florida was enjoined from using the results of its minimum competency test to determine which students would receive high school diplomas, because it had not established the **instructional validity** of its test. At the risk of oversimplifying the ruling, the judge suggested that use of the competency test to deny a high school diploma was not fair because the state could not show that every student had had an opportunity to be instructed in the skills needed to answer the items on the test. Instructional validity is thus concerned with the correspondence between test content and the content of instructional programs. This type of validity issue is likely to become increasingly important as state governments rely on tests to make critical decisions concerning promotion and graduation of public school students.

Assessing Reliability

The **reliability** of a measurement procedure is the technical term for its consistency. It is well known that no measurement procedure—whether it is the use of a bathroom scale or the administration of an achievement test—is perfectly consistent. If you were to step on and off your bathroom scale five times in a row, chances are you would observe three or four different measurements of your weight. Depending on whether you placed your feet in exactly the same spot on the scale, and on the quality of the scale itself, the scale's indication of your weight might fluctuate by three or four pounds across the five measurements. The more consistent the indications of your weight, the more reliable is your bathroom scale. If the scale were to indicate exactly the same weight for you every time you stepped on and off (a highly unlikely occurrence), it would be totally consistent. We would then say that it was perfectly "reliable" *even though it might indicate the wrong weight every time.* (In the *everyday* sense of "reliable" we would include validity, too; but in the technical sense, the validity is a completely separate issue.)

When you observe the weight indicated by your bathroom scale, you are reading what measurement theorists call an **observed score.** This observed score can be thought of as consisting of two parts: one is your true weight (the **true score**), and the other is the result of a variety of factors that have nothing to do with your true weight (the **error component**). These other factors might include such things as the particular position of your feet on the scale (step too far forward, and the scale might read low; step too far backward and it might read high); the fact that you placed the scale on a thick rug instead of on the floor; and the tendency of the scale's springs to fatigue and recover depending on the rates at which you climb on and off the scale. In *every* measurement procedure, regardless of what is being measured or how it is being measured, the observed scores that result from taking a measurement are equal to the sum of the true scores and the error components. The reliability of the measurement procedure depends on the relative sizes of the true scores and the error components. The larger the error components, the lower the reliability; the smaller the error components, the higher the reliability.

Measurement reliability is usually expressed as an index that can take on values between zero to one, much like a correlation coefficient. Unlike a correlation coefficient though, the reliability of a measurement procedure can never be a negative number. A reliability of zero means that the observed scores consist entirely of error components. A reliability of one would mean that the observed scores consist entirely of true scores; in reality, however, observed scores are composed of both true-score and error components, and their reliabilities are less than one.

Another statistic that is closely associated with reliability is called the **standard error of measurement.** The standard error of measurement is equal to *the standard deviation of the error components* in a set of observed scores. If the reliability is equal to one, there are no error components, and the standard error of measurement is equal to zero. If the reliability is equal to zero, observed scores consist totally of error components, and the standard error of measurement is equal to the standard deviation of observed scores.

In many research and evaluation reports that include assessments of measurement quality, standard errors of measurement are provided, together with estimates of the reliability of measurement procedures. These standard errors of measurement can be interpreted by assuming that errors follow a normal distribution, as is commonly the case. For example, if a 50-item test were administered to a group of students and the standard error of measurement were reported to equal two points, you could draw the following conclusions from your knowledge of the normal distribution. Sixty-eight percent of the individuals tested would have true scores that were within one standard error of measurement (two points) of their observed scores. Ninety-five percent of the individuals tested would have true scores that were within two standard errors of measurement (four points) of their observed scores. Alternatively, suppose a given individual had an observed score of 30 on the test. You could then say that the chances were 68 percent that this person's true score was between 28 (one standard error of measurement below 30) and 32 (one standard error of measurement above 30). Also, the chances would be 95 percent that this person's true score was between 26 and 34.

Although reliable measurement procedures are always desirable in research and evaluation, your need for a particularly high level of reliability might depend on the nature of your research, since high reliability may be expensive and unnecessary. And, of course, reliability is not at stake when the true score is changing *on its own.* For example, if you were measuring weights using a bathroom scale, you would want the measurement procedure to provide consistent results at a particular time on a particular day. But you wouldn't be distressed if the scale showed different weights for people on Friday than it did on Monday. Since individuals' true weights change over a period of five days, the scale's indications of their weights should also change. In contrast, suppose that you administered an IQ-test to a group of adults on Monday, and then administered the same test to them on Friday. Ignoring the possibility they might recall the items and hence do better because they've had practice, you would expect the intelligence levels of these individuals to measure exactly the same on Friday as they were on Monday. In this example, reliable measurement implies consistency over a five-day period. Consistency over time is known as **stability** of measurement, and should be high for stable variables.

A variety of methods have been developed for assessing the reliability of a measurement procedure. Some of these procedures indicate consistency of measurement at a given time, and others indicate stability of measurement across a period of time. Whether a particular method for assessing reliability is appropriate depends on the purposes of the research or evaluation study in which it is used. We will now describe several of the most widely used methods of assessing reliability, and discuss the situations in which they are most appropriate.

Suppose you were interested in assessing the reliability of an IQ-test, in the sense of providing stable measurement across a two-week period. One way you could handle this problem would be to administer the IQ-test to a sample selected from the group of individuals you intended to use it with, and then administer the test to these same individuals two weeks later. You could then calculate the correlation coefficient between the scores (or ranks) resulting from the first administration, and the scores resulting from the second. The correlation coefficient you obtained through this procedure would be called a **reliability coefficient.** The higher the correlation coefficient, the higher the correlation coefficient, the higher the reliability of the IQ-test. The logic underlying this **test-retest method** of assessing reliability is as follows: If the measurement procedure is reliable, the person who scored highest on the first administration of the test should score at or very near the top on the second administration. The person who scored second-highest on the first administration should score high (or rank close to second-highest) on the second administration, and so on. In short, with reliable measurement, people ought to retain their relative scores or ranks from one administration of the test to the next. The greater the extent to which this occurs, the higher the reliability coefficient. (Of course, you'd use two different correlation coefficients to calculate the reliability coefficient, depending on whether you correlated scores or ranks.)

The test-retest method of assessing reliability has at least two potential flaws. If it is used with a measurement instrument like an IQ-test, there is always the chance that examinees will either remember the items and do better on them the second time around (the *practice effect*) *or* will remember the *answer* they gave to specific test items on the first administration, and give exactly the same answers on the second administration of the test. This kind of *memory effect* has nothing to do with the *true* reliability of the test, but will tend to increase the correlation between scores obtained on the two administrations. Therefore the *indicated* reliability of the test, when using the correlation between scores as its measure, will be spuriously increased. To handle this problem, publishers of commercially-available tests often create several forms of their tests, so that one form can be administered initially and a second form, containing totally different items, can be used for the second administration. Test forms that are designed to be interchangeable are known as **parallel forms,** and the use of parallel test forms in a test-retest reliability study is known as the **parallel forms method** of assessing reliability.

The parallel forms method of assessing reliability has advantages beyond its elimination of potential memory effects. In most testing situations, a researcher or evaluator is interested in determining an examinee's performance on an entire *domain* of skill, that is, a huge range of test items, rather than the few specific items on one test. A single test merely provides an estimate of how the examinee would perform, by sampling from that huge range. By using two parallel test forms in assessing reliability, the researcher can get a better estimate of the examinees' performances on the entire domain of items simply because more items are used. In other words, the parallel forms method of reliability estimation reduces an important error component, namely the error due to selection of a specific, relatively small, sample of test items. The test-retest method, using a single test form, ignores this error component and might therefore provide an unrealistically high estimate of a test's reliability.

Both the test-retest and parallel forms methods of assessing reliability require that a group of individuals be measured twice. This is costly, time-consuming, and inconvenient. It is often difficult, if not impossible, to assemble the same group more than once. A number of reliability assessment methods that require data to be collected on only one occasion have been developed in response to these problems. We will describe the methods that are used most often.

Reliability assessment methods that require data to be collected on only one occasion are called **internal consistency methods.** Although these methods differ in methodological detail, each of them indicates the degree to which the components of a measurement procedure tend to assess the same underlying variable or construct. When applied to a test, these methods show the extent to which the items in the test tend to measure the same thing.

An easy way to increase reliability is by using a longer test; e.g., give both parallel forms at one sitting. If you wanted to measure a baseball player's hitting ability, you'd get a more reliable measure of this variable by calculating the average percent of times at bat that the player got a hit in both games of a double-header, than by calculating the percentage for a single game. In this example, the two baseball games are analogous to two sets of test items such as two parallel forms. You will obtain far more reliable information about an examinee's performance from twice as many test items administered at one time as you will from half as many, and in general, the greater the number of measurements involved in a measurement procedure, the more reliable the procedure. So the longer a test—up to the point where examinee fatigue sets in— the more reliable it will be *provided* the added measurements assess the same variable as the initial ones. If you were trying to assess a ballplayer's hitting ability, you would not increase measurement reliability by counting the number of times he or she caught a fly ball. Fielding skill doesn't correlate highly with batting skill. The added measurements would have to assess hitting ability, just like the initial ones. Internal consistency reliability methods indicate whether, and to what degree, all of the measurements used in a procedure assess the same variable.

One of the simplest internal consistency methods for assessing reliability is called the **split-half method** or the **odd-even method.** In this procedure, we *make* two tests out of one. Thus, a group of examinees is given a test only once, but two scores are calculated for each examinee, one based on the odd-numbered items in the test, and the other based on the even-numbered items. These two sets of scores are then correlated, to determine whether examinees who score highest on the odd-numbered items also tend to score highest on the even-numbered items. This correlation coefficient is called the **split-half reliability coefficient.** The higher the split-half reliability coefficient, the more consistent are the items on the test (in the sense that the odd-numbered items measure the same variable as the even-numbered items).

Because the reliability of a measurement procedure increases as the number of measurements is increased, the split-half reliability coefficient will tend to underestimate reliability. For example, if it is applied to a 50-item test, the split-half method really only indicates the reliability of two shorter tests—the 25-item halves, composed of the odd items and the even items, respectively. To estimate the reliability of the full 50-item test—higher because it is longer—it is necessary to adjust the split-half reliability coefficient using a formula known as the **Spearman-Brown prophecy formula.** When you see a reliability coefficient reported as a Spearman-Brown coefficient, you can be sure that the author has used the split-half method for assessing reliability, and then has applied the adjustment formula.

The split-half method of assessing reliability, like all internal consistency methods, will not show the stability of a measurement procedure over time. Nor will it reflect errors that arise from peculiarities in the administration of a measurement procedure on a particular occasion. It merely picks up some of the error due to the limited size of the sample of particular measurements or test items used and may overestimate the true reliability of a measurement procedure because it does not pick up other possible errors such as excessive help given by the tester.

Two widely used—but not necessarily superior—alternatives to the split-half method of estimating reliability are named after the authors of the research paper in which they were first proposed. Both methods can be used only with procedures composed of a series of measurements that are scored "one" for a correct response and "zero" for an incorrect response. These procedures are known as the **Kuder-Richardson Formula 20 method** and the **Kuder-Richardson Formula 21 method.** You will often find them

abbreviated as KR-20 and KR-21. The KR-21 formula is sometimes used instead of KR-20, because it is simpler and easier to
compute. However, it makes the assumption that each of the
measurements that compose a procedure (or each of the items
that compose a test) is equally difficult. This assumption is often
violated in practice, resulting in an overestimate of true
reliability.

The Kuder-Richardson methods of assessing reliability have all
of the advantages and all of the pitfalls of the split-half method.
They are sensitive to the same kinds of measurement errors, and
ignore the same error components as well. They should never be
used to assess stability of measurement, nor in situations where
the conditions of measurement are difficult to control nor where
speed of response is a major factor in performance. In any of
these applications, they are likely to give a falsely inflated estimate of true measurement reliability.

It is not necessary to use the Spearman-Brown Prophecy
Formula to adjust the reliability estimates obtained through use
of the Kuder-Richardson procedures, as must be done with the
split-half method. The correction for reduced test length is built
into the KR formulas.

A final method for assessing internal consistency reliability
was named after its developer, Lee J. Cronbach. It is known as
Cronbach's Alpha. Cronbach's procedure, like the other internal
consistency methods, uses measurement data collected on a
single occasion. The alpha method is a generalization of Kuder-
Richardson Formula 20, in that the test items (or other components of the measurement procedure) do not have to be scored
either zero or one. This improvement makes it possible to apply
Cronbach's method to many measurement procedures other
than tests, including attitude instruments in which each item
requires a response on a five-point scale that might range from
"Strongly Disagree" to "Strongly Agree."

In recent years, Cronbach's Alpha has been used in more
sophisticated studies of reliability known as **generalizability
studies.** In these studies, the components of measurement error—such as those due to instability and those due to peculiarities of measurement conditions—are analyzed separately. For
example, in a study in which five judges were used to estimate
the work attitudes of a group of employees on three different

occasions in two different work settings, generalizability methods would make it possible to determine how much the reliability was affected by disagreements between judges, and how much it was affected by changes across time, or by inconsistency across work situations. This kind of information could then be used to design a more-reliable measurement procedure.

Reporting the Scores

The score scales that are used in research in education and the behavioral sciences are often arbitrary or difficult to interpret. For example, suppose that a researcher used a 20-item attitude scale in which each item provided five responses ranging from "Strongly Disagree" to "Strongly Agree." It is traditional to give a score of one for a response of Strongly Disagree, and a score of five for a response of Strongly Agree. With 20 items, a respondent could earn a score that ranged from a low of 20, to a high of 100. This score scale is totally arbitrary. It would make just as much sense to give a score of two to a Strongly Disagree response, and a score of 10 to a Strongly Agree response. In that case, total scores would range from a low of 40 to a high of 200. If the original scoring system were used, but the attitude scale had 25 items instead of 20, the possible scores would range from a low of 25 to a high of 125. In all three cases, we could presume that a higher score on the scale indicated a stronger level of agreement. However, any numerical value on the score scale would be an artifact of the way the scale was constructed, and would have no absolute meaning.

This example illustrates the nature of observed scores that result in many research and evaluation studies in education and the behavioral sciences. To give meaning to the scores, it is necessary to transform them in some way. Various transformations have been developed for this purpose.

One way to give meaning to a person's raw score is through comparison with the scores earned by others. This is called a **normative interpretation** of scores. We discussed one such transformation, the percentile rank, in Chapter 1. If a person earned a score of 75 on our 20-item attitude scale, we could judge the relative value of this score by computing the person's percentile rank. For example, if a raw score of 75 corresponded to a percentile rank of 80, we could conclude that the degree of the person's agreement was at least as strong as that held by 80 percent of the other individuals who were assessed.

As we mentioned in Chapter 1, percentile ranks are group dependent. A given score might have a relatively high percentile rank in one group, and a relatively low percentile rank in another. Publishers of commercially-available tests often report the "national" percentile rank associated with every possible raw score on their tests. These precentile ranks purpose to indicate the percentage of examinees at a particular grade level or of a particular age throughout the nation, who would earn scores that were equal to or less than each possible raw score. Percentile rank distributions of this kind are called **national norms.** In recent years, test publishers have encountered increasing difficulty in obtaining truly representative samples of examinees when developing their national norms. As a result, the samples used vary substantially from one test to the next, and so-called national norms are often not nationally representative, nor comparable between different tests.

Another type of normative transformation results in what are termed **standard scores.** A standard score is a score that is on a scale with a pre-defined mean and a pre-defined standard deviation, selected for computational and interpretational convenience. Several types of standard scores are commonly used in educational and behavioral science research. They differ only in the definition of their means and standard deviations.

The most basic standard score is termed a **z-score.** The lower-case "z" is used here to denote a score scale with a mean value of zero and a standard deviation of 1.0. If a set of raw scores is transformed to z-scores, a raw score that is right at the mean will have a corresponding z-score of zero; a raw score that is one standard deviation above the mean will have a corresponding z-score of 1.0; and a raw score that is two standard deviations below the mean will have a corresponding z-score of minus 2.0.

One advantage of z-scores is that they allow comparison of an individual's relative performances on several different measures, even though the measures have totally different raw-score scales. For example, if a student scored 25 on a 40-item reading test, and 30 on a 50-item math test, we could not tell from these scores whether that student's performance was better in reading or math. However, if the student's z-score was 1.0 on the reading test and 1.5 on the math test, we could confidently conclude that, compared to the other students with whom he or she was tested, the student was doing substantially better in math than in reading. We've "built in" all the information about means and standard deviations on the two tests, into the z-scores themselves.

Another popular standard score scale has a mean of 50 and a standard deviation of 10. Scores on this scale are called **Z-scores** or **T-scores.** A person with a Z-score of 50 would be right at the mean of his or her group, and a person with a Z-score of 70 would be two standard deviations above the mean or his or her group. Similarly, a person with a Z-score of 20 would be three standard deviations below the mean or his or her group, since 20 is 3×10 or 30 points below the mean of 50.

Historically, T-scores were computed in such a way that their distribution was always normal in shape, whereas the shape of a distribution of Z-scores was identical to the shape of the distribution of raw scores from which they were computed. In recent years, this distinction between T-scores and Z-scores has been abandoned, and the two terms are used interchangeably.

Yet another standard-score scale was developed by the College Entrance Examination Board (CEEB) for reporting scores on its college entrance tests, including the *Scholastic Aptitude Tests (SAT)* and the *Graduate Record Examination (GRE).* Originally, the CEEB scale had a mean of 500 and a standard deviation of 100. So a person who earned a score of 600 on one of the parts of the SAT or GRE would be one standard deviation above the mean of the distribution of those who completed that part of the test. Since examinees rarely score more than three standard deviations below the mean or more than three standard deviations above the mean of a distribution, virtually all examinees would earn CEEB scores between 200 and 800. For that reason, the College Entrance Examination Board reports scores that are outside the 200-to-800 range as being *at* the end points, and thus reports all scores as being in the 200-to-800 range.

When large samples of individuals are measured or tested, the resulting distribution of scores often follows the normal distribution. From our discussion of the normal distribution in Chapter 3, you might recall that specific percentages of the distribution fall within an interval that is one, two, or three standard deviations on each side of the mean. These facts can be put to good use in establishing relationships between percentile ranks and various standard-score scales, whenever scores are normally distributed. In 1955, the Psychological Corporation published a graph that summarizes many of these relationships. It is reproduced below as Figure 1.

Figure 1 shows the percentages of scores in a normal distribution that fall within score intervals that are one standard deviation wide, on each side of the mean of the distribution. For example, 13.59% of the scores will fall between score values that are one standard deviation above the mean and two standard deviations above the mean. Scores that are at the mean, or one, two, three, or four standard deviations above or below the mean, are shown for a number of standard score scales, including the z-score scale, the T-score scale, and the CEEB score scale that have already been discussed.

Other standard score scales shown in Figure 1 include the AGCT scale, the stanine scale, the Wechsler scale for subtests, and the Deviation IQ scale. We will discuss each of these briefly. The AGCT scale was developed by the U.S. Department of Defense during World War II, for reporting the performance of military personnel on the Army General Classification Test (AGCT). The scale had a mean of 100 and a standard deviation of 20. Both the scale and the test have been superseded by the Armed Services Vocational Aptitude Battery (ASVAB), and a percentile rank scale. The word "stanine" is a contraction of the words standard and nine. This score scale was developed by the U.S. Air Force, but is now widely used by publishers of achievement tests intended for use in elementary and secondary schools. The scale has nine values—the whole numbers one through nine. Each stanine from two through eight represents an interval on the score scale that is one-half a standard deviation wide. Stanine 5 is located around the mean, in the middle of the score distribution, and includes scores that are between one-fourth of a standard deviation below the mean and one-fourth of a standard deviation above the mean. Stanines 6 through 8 are located right above Stanine 5, and are also half a standard deviation wide. Stanine 9 is at the top of the distribution, and includes any score that is at least one and three-fourths of a standard deviation above the mean. Therefore, any raw score that has a correspond-

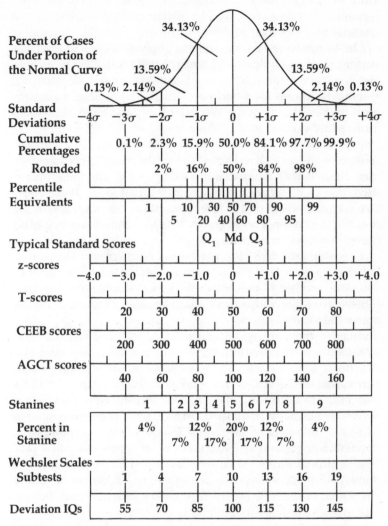

Figure 1. Typical Standard Scores Related to the Normal Curve*

*The Psychological Corporation. Test Service Bulletin No. 48, January 1955. Printed by permission of the publisher, Psychological Corporation.

ing z-score of at least 1.75 is in Stanine 9. Stanines 1 through 4 are located below Stanine 5, and have definitions that are similar to those of Stanines 6 through 9.

The **Wechsler** scales for subtests apply to several of the most widely-used individually-administered IQ tests. These include the Wechsler Adult Intelligence Scale (WAIS) and the Wechsler Intelligence Scale for Children (WISC). The scale is used with each of the subtests of the Wechsler tests, and has a mean of 10 and a standard deviation of three.

The **deviation-IQ** scale is used with virtually all group-administered IQ tests. It is defined so that the mean IQ of examinees in the publisher's national norm group is 100, and the standard deviation of IQ scores in this group is equal to 15. Therefore, a person with an IQ of 115 is one standard deviation above the mean of the national norm group. Provided the distribution of IQ scores is normal in shape, a person with an IQ of 115 would have a percentile rank of 84 in the national norm group. This is true because 50 percent of the group would score below the mean of 100, and another 34 percent of the group would score between the mean of 100 and a score that was one standard deviation above the mean (115).

Because they are widely used in educational research and evaluation studies, two additional score transformations must be described. One is a standard-score scale called the **normal curve equivalent (NCE)** scale. This score scale has a mean of 50 and a standard deviation of 21.06. Its only advantage is that, when scores are normally distributed, an NCE score of one corresponds to a percentile rank of one, an NCE score of 50 corresponds to a percentile of 50, and an NCE score of 99 corresponds to a percentile rank of 99. The scale was developed for the U.S. Department of Education for use in evaluating education programs supported under Title I of the Elementary and Secondary Education Act. As federal requirements for evaluation of education programs are reduced, it is likely that the NCE scale will fade from the scene.

The **grade equivalent scale (GE)** is, without doubt, the most widely used score transformation in educational research and evaluation studies. Its seeming, but deceptive, simplicity is responsible for its overwhelming popularity. The scale is based on a division of the school year into ten equal units, each representing a "month." Since most school years are nine months long, the three summer months are regarded as equivalent to one month of schooling.

When an achievement test is administered and a student is reported to be scoring at grade equivalent 4.7, that student's performance, regardless of his or her actual grade level, is supposedly equivalent to the median performance of students who are beginning the seventh month of their fourth school year. Likewise, a grade equivalent score of 6.0 supposedly indicates performance that is equivalent to the median of the score distribution of students who are just beginning their sixth year of school. The score scale appears to be simple and straightforward.

The problems with grade equivalent scores arise from the way they are constructed. Typically, a test publisher will administer an achievement test intended for students at a particular grade level to a sample of students in that grade, and to samples of students one grade above and one grade below that grade level. Suppose that the test is administered to students in grades four, five and six, during the eighth month of the school year. The median score earned by fourth-graders is then set equal to a grade equivalent score of 4.8; the median score of fifth-graders is GE 5.8; and the median score of sixth-graders is GE 6.8. These three score values are then used by the publisher to specify all other grade equivalent scores. A smooth curve that passes through the three points is used to define grade equivalent scores between 4.8 and 6.8, and to extend the scale below 4.8 and above 6.8. The process is subjective, assumes a good deal, and is liable to be very misleading.

Grade equivalent scores should be avoided whenever reasonable alternatives are available, and used with caution when they cannot be avoided. You should realize that a third-grader who earns a grade equivalent of 5.2 on a third-grade test in reading is not thereby proven ready for a fifth-grade reading group. He or she merely did *as well as some fifth-graders on a third-grade test*. Furthermore, grade equivalent scores are *not* on an interval scale. They are merely on an ordinal scale, so they should not be averaged, nor used to compute standard scores. Grade equivalent scores on one test are not equivalent to grade equivalent scores on another. Therefore, you cannot safely conclude that a student with a grade equivalent of 6.0 on a reading test and a grade equivalent of 5.5 on a math test is doing better in reading than in math.

Summary

Measurement is an essential component of research and evaluation in education and the behavioral sciences. In this chapter we have briefly described the fundamental measurement issues of validity, reliability, and scales for reporting scores. Since many of the variables that are measured in education and the behavioral sciences cannot be observed directly, the correspondence between the score that results from some measurement procedure and a person's status on an underlying variable must always be questioned. The validity of a researcher's interpretations of measurement is always a troublesome issue. The validity issues that arise in various research and evaluation contexts were discussed under the headings of construct validity, concurrent validity, predictive validity, and content validity.

Consistency of measurement, termed measurement reliability, is another essential feature of a good measurement procedure. In addition to defining reliability in terms of stability of measurement and internal consistency, we described the features of a number of procedures for estimating reliability. The test-retest method and the parallel forms method provide estimates of stability of measurement. The split-half method, the Kuder-Richardson formulas, and Cronbach's alpha provide estimates of internal consistency reliability.

Since most raw-score scales used in education and behavioral sciences research are arbitrary, score transformations are useful in order to interpret results in these fields. The score transformations described in this chapter include percentile ranks, grade equivalent scores, z-scores, T-scores, Deviation-IQ scores, and stanines. The relation of these (and other) score scales to each other and to the normal distribution was discussed or displayed graphically.

Measurement is an important and complex topic that has given rise to year-long graduate courses in universities throughout the nation, and text material that encompasses many volumes. In this brief chapter we have merely "hit the highlights." But by paying careful attention to the fundamental issues of measurement validity, reliability, and scales of measurement discussed here, you can become a much more sophisticated and better-informed consumer of research and evaluation in your field. We urge you to question the results of measurement and accept authors' conclusions with appropriate caution and scepticism.

Problems—Chapter 5

5.1 You are considering the adoption of a new reading test for your school system. Publisher A lists the test-retest reliability of its reading test as 0.87. Publisher B lists the Kuder-Richardson Formula 20 (KR-20) reliability of its reading test as 0.91. If you are seeking the reading test that will provide the most stable measurement of reading skill over a specified time interval, whose test should you adopt?

    a. Publisher A's because its reliability was estimated using the test-retest method.
    b. Publisher B's because the reliability of its test is highest.
    c. Publisher B's because its reliability was estimated using the KR-20 method.
    d. It doesn't matter, because both tests have high reliability.

5.2 Publisher A reports performance on its mathematics achievement test in terms of the percentage of objectives each student has "mastered." Publisher B reports performance on its mathematics achievement test in terms of the percentile rank of each tested student. If you were to characterize the tests offered by these publishers as either norm-referenced or criterion-referenced tests, which of the following would most likely be an accurate characterization:

    a. The tests of both Publisher A and Publisher B are most likely norm-referenced.
    b. Publisher A's test is most likely criterion-referenced, and Publisher B's test is most likely norm-referenced.
    c. Publisher A's test is most likely norm-referenced and Publisher B's test is most likely criterion-referenced.
    d. The tests of both Publisher A and Publisher B are most likely criterion-referenced.

5.3 You are interested in adopting a test that will help in selecting students who are most likely to do well in an honors high school physics program. In reviewing the technical literature on available commercial tests, which of the following pieces of information should most influence your decision on the test to be adopted?
   a. A widely-used science aptitude test has a reported test-retest reliability of 0.85.
   b. A newly-developed achievement test in high school physics has a reported concurrent validity of 0.90 when used in conjunction with a widely-accepted test of high school physics achievement.
   c. A well-recognized science aptitude test has a reported predictive validity of 0.87 when a widely-accepted test of high school physics achievement is used as a criterion measure.
   d. A well-accepted test of high school physics achievement was found to have very high content validity when compared to the objectives of your honors physics curriculum by a group of science curriculum specialists.

5.4 Suppose that you wanted to explain the standardized test performances of students named Bob, Mary, and Joe when you held a conference with their parents. Bob's performance on a standardized English test gave him a T-score of 70; Mary's performance on a standardized Social Studies test gave her a percentile rank of 87; and Joe's performance on a standardized Mathematics test gave him a z-score of 0.25. All of these scores involve national norms. Which of the following interpretations would be most accurate?
   a. Bob and Mary performed well above average, compared to the national norm group, but Joe's performance was close to the national average.
   b. All three students performed well above average compared to the national norm group.
   c. Mary's performance was well above the average of the national norm group; Bob's performance would rate him about a "C" on the usual grading scale; and Joe's performance was just below average.
   d. In terms of their performance relative to each other, Joe did best, Mary did next best, and Bob did least well.

6.

Elements of Statistical Inference

Inferential Statistics

A busy administrator—for example—is often faced with information in the form of numbers that he or she must use in making important decisions, and sometimes the decision to be made requires *going beyond the numbers presented* to answer a series of "what if" or "is it also true" questions such as "what will happen to class size if we wait one more year before adding extra teachers?". In going beyond available data, one must use a form of **inference,** and because the numbers involved are statistics we call this statistical inference. Here are some more examples of what we mean, put in terms of an administrator's problems.

1. A school system up-state used high-achieving sixth-graders to tutor struggling third-graders in reading. You've heard through the grapevine that it worked very well for them. How well did it really work? And is it likely to be effective in your school system? You'll have to look carefully at the data to answer even the first question, which calls for an evaluation. And then you'll have to *go beyond the data* to infer what would happen if you tried the approach in your school system.

2. At the last meeting of the American Association of School Administrators, you heard a talk on the relationship between good teacher attitudes and student learning. The speaker claimed that teachers with better attitudes were clearer and better organized in the classroom, and that clarity and organization on the part of teachers resulted in better achievement on the part of students. Just how strong were those relationships? What kinds

of teacher attitudes seemed to make a difference? Should you invest in motivational workshops for teachers? Better look at the data very carefully! And then you'll have to **infer** from the data to your own situation.

3. Although career education is no longer the darling of bold-thinking educational innovators, many still feel that high school students leave the twelfth grade with insufficient information on the world of work. They feel that these students have a vague understanding, at best, of the relationship between education and career opportunities. In your school system, the career education program disappeared in 1978, along with performance contracting, new math, and strict accountability. In your view, some educational fads are best sent to an early grave, but others are abandoned too quickly, without ever getting a fair test. You'd like to see greater emphasis on the transition between school and work in your district's high school curriculum, and you've just been reading about an apparently successful career knowledge development program that was used in a school system in Pennsylvania. Once again, the old questions surface. How well did the program work in the Pennsylvania school system? You read that eleventh-graders who had been in the program a year knew more about career opportunities than did a similar group of students in schools who hadn't participated in the project. Did the project students' increased knowledge reflect a real project effect, or was it just due to the selection of especially bright students as project participants? Did the project students really know a lot more about careers at the end of the school year than at the beginning? Did they learn more than you'd expect them to, just by growing a year older and going to school a year longer, even without the project? How sure can you be that the students' apparent gains in knowledge wouldn't just vanish, were you to develop a similar project in your school system? Once again, these questions call for a hard look at relevant data. They also require inference beyond the data to a new situation or a new group of students.

All of us make inferences in our everyday lives. Sometimes we just play our hunches (Ms. Jones has always supported me on school board budget votes in the past; I assume she'll vote for my budget proposal again this year.). In other cases, we're more systematic in gathering information and in generalizing from that information (I don't know how parents in my school system feel about our physical education program. I think we'll do a survey to get some information on parental opinions.). Regardless of the degree of formality or rigor you use in forming your inferences, as an administrator paid to make decisions and implement policy, you can't get away from the necessity of going beyond the facts at hand. You must infer the likely actions of people, the likely consequences of your decisions, and the probable shape of the future in a world fraught with uncertainties.

This section of the book will teach you how statistics can be used in making formal inferences from numerical data. As in the first section, we will emphasize how to read and understand evaluation and research reports, and how to figure out what they do and don't say. We assume that you're too busy to compute your own statistics, and that you're not interested in learning a lot of fancy formulas. We assume that you are interested in knowing how to make sense out of all those numbers in evaluation and research reports. In fact, as an administrator, you've got to be able to read and understand evaluation reports just to survive. The days of "flying by the seat of one's pants" are over. All administrators face an angry public that demands more effective schools, smaller school budgets, and generally increased efficiency. The only way to meet those demands is through careful selection, development, and adoption of instructional, operational, and personnel strategies that are effective, economical and efficient. Evaluation can show the way to do that, and statistical inference is an essential part of many evaluation problems.

Our Approach

Every field has its own vocabulary and its own set of basic concepts; inferential statistics is no exception. So we'll begin at the beginning, with some basic vocabulary and the fundamental ideas that are used in all formal statistical inference. That will be the dry part, where you'll wonder what inferential statistics has to do with becoming a more effective decision-maker, and how it will help you understand evaluation and research reports; but it's only a couple of pages. Once we get through the basics, we'll move immediately to applications. You'll find that the same logic is used repeatedly in all applications of inferential statistics. So spending a good bit of time and effort on the next few chapters will pay off in the end. Once we get beyond fundamental concepts, every application of inferential statistics will be illustrated with data from a real evaluation report. You'll learn to read and understand the data tables, to form your own conclusions on what the data say, and to recognize when an author is handing you a line instead of the straight facts. Flushes of programmatic enthusiasm abound in evaluation reports, and you've got to be a knowledgeable reader to avoid getting bushwhacked. You don't want to buy into a major curriculum change or in-service training program just because some over-zealous evaluator took poetic license with his or her data.

Some Basic Vocabulary

Two things go on in statistical inference: **estimation** and **hypothesis testing.** We either estimate, or test hypotheses about various aspects—which we call **parameters**—of **populations.** In doing so, we use data collected from **samples** to compute various **statistics.** When we're doing estimation, the statistics we compute are called **estimates.** The formulas we use to compute them are called **estimators.** Now let's define all of these terms.

Population. Formally, a population is any collection of objects or entities that have at least one characteristic in common. Although we usually think of populations of people, in statistical work our populations can consist of schools, school systems, governmental units, pieces of furniture, library books, or any describable group of things or persons that share some common characteristic. In all of the inferential situations we describe later in the book, you will see that the population is that group of objects or persons about which we want to make some inference. That is, we really want to know some characteristic of a population, even though we haven't measured or observed every person or object that makes up the population.

Parameter. A parameter is some characteristic of a population. For example, if the population we're interested in consists of all fourth-graders enrolled in our school system, one parameter of that population is the fourth-graders' average IQ test score. Another parameter is the percent of fourth-graders who need remedial reading instruction. In statistical work, parameters are numbers that describe some important features of a population.

Unless we've measured or observed every object or person in the population we're interested in, in every important respect, we won't know the true value of the population's parameters. So we use inferential statistics to infer the values of population parameters.

Sample. The simplest definition of a sample is that it is just a part of a population. A sample can consist of an entire population or any part of a population. In all statistical problems, a sample will consist of those objects or people we have observed or measured. For example, if the population we were interested in was all fourth-graders in our school system, and the parameter we were interested in was the average IQ test score of all fourth-graders, our sample would consist of the fourth-graders who actually completed an IQ test. They would be the sample of persons who had supplied us with data.

Many samples are useless for statistical work. We want to do our best to avoid such samples. Think about the problem of determining the average IQ test score of all fourth-graders in your school system. Suppose that you had given an IQ test to a sample of fourth-graders who were enrolled in a special program for gifted students. Their average IQ might be around 120. You wouldn't want to use their average IQ to infer the average IQ of all fourth-graders in your school system. Clearly, your estimate would be too high. The problem is that the fourth-graders in your program for the gifted don't fairly **represent** all fourth-graders in your school system. In statistical work, we only want to use samples that are representative of the populations we're interested in. Such samples are called **probability samples,** and have the following features in common. First, every person or object that is a member of the population has some chance (but not necessarily the same chance) of being a member of the sample. Second, before we select our sample, we can figure out the chances that any person or object, or group of persons or objects, might wind up in the sample. Third, we have defined our population so well that we can state without any question, whether a particular person or object is or is not a member of the population. For the moment, don't worry too much about the second and third features of probability samples. The first feature is often the stumbling block when samples don't represent populations. In the example we've been considering, if all sampled students were in the program for the gifted, then fourth-graders who were not in that program wouldn't have any chance at all of being sampled. We therefore wouldn't have a probability sample, and we'd run into real trouble when we tried to infer the average IQ of all fourth-graders in the school system.

Statistic. A statistic is just a characteristic of a sample. It is a number that describes some feature of a sample (such as the average IQ test score of a sample of fourth-grade students). A statistic can be used to infer the value of a population parameter. You've been introduced to lots of different statistics in the first part of this book. Means, standard deviations, ranges, semi-interquartile ranges, variances, medians, and correlation coefficients are all examples of statistics. Each is computed using the data collected on a sample. Each describes some feature or characteristic of a sample. The statistic can be exactly calculated; there's no risky inference involved.

So far, we've defined four terms: **population, parameter, sample** and **statistic.** Just remember that the two terms that start with the letter "P" go together, and the two terms that start with the letter "S" go together. Parameters describe characteristics or features of populations. Statistics describe characteristics or features of samples. They often describe the same feature. For example, the average IQ test score of a sample of fourth-graders is a statistic; it's *calculated exactly.* The average IQ test score of the population from which the sample was selected is a parameter: it's what we usually have to *estimate.*

Estimate. An estimate is just a number. It is computed by using data collected from a sample, but it is *used* as a "best guess" about the value of a population parameter. Going back to the example we've used before, if you wanted to know the average IQ test score of the fourth-graders in your school system, you might select a probability sample of, say, 100 of them, and administer an IQ test to all of the sampled students. If you computed their average IQ score, and it turned out to be 105.4, that number would be an estimate of the average IQ test score of all fourth-graders in the school system (the population parameter of interest). That estimate is only *probably* correct; that's where the inference of inferential statistics comes in.

Estimator. An estimator is a formula used to compute an estimate. In the example just discussed, the estimator would be the formula for computing a sample average—add up the IQ test scores for each sampled fourth-grader, and divide by the number of sampled fourth-graders.

Problems—Chapter 6

6.1 In a study of the relationship between investment and pay-off, an economist wanted to estimate the correlation coefficient between current price and current dividend rate for all utilities stocks listed on the New York stock exchange. If the list of such stocks was considered to be a population, what would the correlation coefficient be called?

    a. A statistic.
    b. An estimate.
    c. A parameter.
    d. A sample.

6.2 When presidential election time rolls around, major polling companies, like the Harris Poll and the Gallup Poll, often provide predictions of election outcomes. They base their predictions on surveys of about 1500 prospective voters. The 1500 voters they survey constitute:

    a. A population.
    b. A parameter.
    c. An estimator.
    d. A sample.

6.3 When Harris Poll or Gallup Poll researchers gather data from potential voters on how they intend to vote in an upcoming presidential election, the researchers usually summarize their findings by computing the percentage of voters who said they would vote for Candidate A, the percentage who said they would vote for Candidate B, etc. These percentages are examples of:

    a. Population parameters.
    b. Sample statistics.
    c. Population estimators.
    d. Sample parameters.

6.4   If a school superintendent wanted to know the mean IQ-test score of all sixth-grade students in a school system, it would not be necessary to test every sixth-grader. The population mean could be estimated by testing a sample of sixth-graders, and then using the mean IQ-score of the sample to estimate the mean that would have been found, had all sixth-graders in the system been tested. The sample mean would be computed by summing the IQ-test scores of all tested sixth-graders, and then dividing the sum by the number of tested sixth-graders. This formula, used to compute the sample mean, is an example of:

    a. An estimator.
    b. A parameter.
    c. An estimate.
    d. A statistic.

7.

Nuts and Bolts
of Estimation

Estimation is the process of using data from a sample to infer the value of a population parameter. Like all forms of statistical inference, estimation involves the elements defined in the last chapter. The process of estimation is as follows:

1. The population that is of interest is carefully defined. It's important to realize that the way a population is defined will have a lot to do with the outcome of an evaluation or research study.

2. The relevant population parameter (or parameters) is defined. That is, we identify what we want to estimate. Examples include the one we've used earlier (the average IQ-test score of all fourth-graders in a school system), the proportion of windows in a district's schools that need to be replaced, the average and range of annual salaries of the school system's administrative staff members, etc., etc.

3. A probability sampling procedure is used to select a representative sample of the population of interest. In most of the statistical work you'll encounter in evaluation and research reports, a particular probability sampling procedure, **simple random sampling,** will have been used. With simple random sampling, each member of a population has exactly the same chance of being sampled. In addition, the chance that a given member of the population is sampled doesn't depend at all on what other members of the population have been or will be sampled. In practice, simple random sampling is often assumed to have been used, regardless of the actual method used for selecting objects or persons for collection of data. Sometimes the assumption is ap-

propriate, and inferential statistical procedures work well. On other occasions, the assumption is totally unwarranted, and inferential statistics will produce badly distorted results. In a later section, we'll return to a discussion of this pitfall, and how to look out for it.

4. Data are collected (either through observation, measurement, or interview) from each member of the sample. The data collected are appropriate for estimating the population parameter (or parameters) that are of interest. If we were interested in the average IQ-test score of a population, we'd give each member of the sample an IQ test. If we were interested in the proportion of windows in the district's public schools that needed replacing, we'd record, for each sampled window, whether or not it needed replacing, and so on.

5. An appropriate estimator (formula) is applied to the data collected from the sample, to compute an estimate of the population parameter(s) in which we're interested.

The entire estimation process is illustrated by the drawing in Figure 1.

Two Types of Estimation

Two approaches to the estimation of population parameters are used in evaluation and research reports. The first, called **point estimation**, has been described in the preceding section. In point estimation, a population parameter is estimated using a single sample statistic. For example, a sample average (a statistic) is used to estimate a population average (a parameter). Or a sample standard deviation (another statistic) is used to estimate the value of a population standard deviation (another population parameter). In each case, a population parameter is estimated using a single estimate (called a point value), that is computed using the data from an observed or measured sample.

The other kind of estimation is called **interval estimation**. It involves the computation of **confidence intervals**. Regardless of the type of estimation used, the researcher or evaluator is trying to answer the question, "What is the value of a population parameter?" When point estimation is used, the form of the answer is, "I don't know for sure, but my best guess is (the value of the sample statistic used as an estimate)." When interval estimation is used, the form of the answer is something like "I don't know the exact answer, but I'm 95 percent confident that the value falls between the following two numbers." The two numbers that are given in the interval-estimation answer are called a **lower confidence limit** and an **upper confidence limit,** respectively. The interval between the lower confidence limit (LCL) and the upper confidence limit (UCL) is called a **confidence interval.** Confidence intervals are widely misinterpreted in evaluation and research reports. It is important that you understand what they are and what they are not, so that you won't be misled. The explanation must, of necessity, be a little bit involved. At times it might read like a shaggy dog story and you may wonder whether it will ever end. But stick with it. Once you understand confidence intervals, you'll know a good bit about the logic of statistical inference. Many data tables in many evaluation reports will then seem a lot clearer.

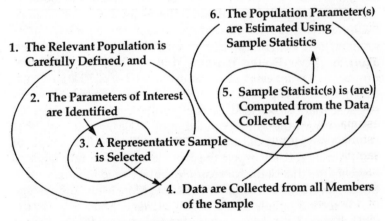

Figure 1. The Process of Estimation

To explain confidence intervals, we'll have to engage in a hypothetical discussion. It's important that you understand right off that the discussion is purely hypothetical. It's merely a teaching device, and won't have anything to do with actual practice. To keep fiction and reality clearly separated, let's briefly review the actual steps involved in estimating a population parameter. You'd begin by defining the population you were interested in and the parameter you wanted to estimate. Then you'd select a *single* probability sample from that population. After you'd collected data from all members of the sample, you'd compute an estimate of the population parameter. The estimate could be in the form of a single value (a point estimate) or in the form of a confidence interval (an interval estimate).

Suppose that you've identified the population you're interested in and the population parameter you want to estimate. For the sake of specificity, let's go back to the example we've used previously, and assume that you want to estimate the average IQ score of all fourth-grade students enrolled in the public schools in your district. Let's assume that you're an administrator in a fairly large school district, and that you have 5000 fourth-graders in your population. Now you wouldn't have to test all 5000 students in order to estimate their average IQ score. In fact, it might be sufficient to administer an IQ test to a fairly small sample (say 100 students) and use their IQ-test performance to estimate the average IQ of the entire population. Here's the hypothetical part. Suppose that you selected a simple random sample of 100 fourth-graders, gave each one an IQ test, and when you computed their average IQ you found it equalled 105.2. You could use that number as a point estimate of the average IQ of all 5000 students, but in this case you want to form a confidence interval. Now let's suppose that you selected a second simple random sample of 100 students from the same population of 5000. Do you think that their average IQ would equal 105.2, just like the first sample of 100? You'd probably expect the average for the second sample to be similar to that of the first sample, but it would be surprising if the two averages were exactly the same. More likely, the average for the second sample would be a bit higher or a bit lower. For the sake of specificity, let's assume that the average for the second sample was 101.0.

Proceeding with this hypothetical example, let's assume that you were to select a large number of independent simple random samples from the population of 5000 fourth-graders. Each random sample would contain exactly 100 students, and after you had given all students in a sample an IQ test, you'd record their average IQ-test score, and you'd put them back in the population, so you wouldn't run out of students to sample. Remember, this is all hypothetical, just to illustrate the logic and meaning of

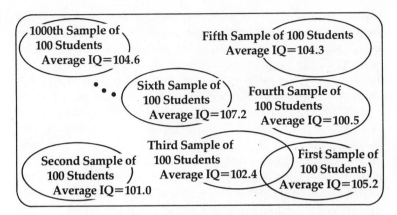

Figure 2. **Hypothetical Example of Repeated Sampling and Estimation From a Population of 5000 Fourth-Grade Students.**

confidence intervals.

At this point, you'd have a long list of sample average IQ's, one for each sample you selected. If you were to select 1000 samples, each with 100 students, before you got tired of the whole process you'd have a list of 1000 sample average IQ's. Look at Figure 2 for an illustration of the results of this process.

From known statistical theory, we can tell what would happen if you were to treat the 1000 sample averages just like any set of data, and were to draw a frequency distribution of the 1000 scores. The shape of the frequency distribution would be a familiar bell-shaped curve, called a **normal distribution.** Furthermore, we know that the average of all the 1000 sample averages would be exactly the same as the average IQ of the entire population of 5000 fourth-grade students. This is a very important point. It tells us that the average of all of our estimates would exactly equal the population parameter we're trying to estimate. (Statisticians pride themselves in being right, "on the average.") Finally, if we were to compute the standard deviation of the 1000 sample averages, we would find that it was related to the standard deviation of the individual IQ scores of the 5000 fourth-graders who make up the population. In this case, the standard deviation of the 1000 sample averages would be one-tenth as large as the standard deviation of the individual IQ scores of those 5000 students. That's because each of our samples contains 100 students. In general, the standard deviation of the sample averages would equal the standard deviation of the original scores, divided by the square root of the sample size. If that last bit of information is a bit mind boggling, don't worry about it right now. Just accept the information concerning this example on faith. Figure 3 shows what the frequency distribution of the 1000 sample averages would look like. To make the problem more concrete, we've assumed that the standard deviation of the individual IQ-test scores of the 5000 students in the population equals 15 points. That's close to the true value we'd find with most standardized IQ tests. Since the standard deviation of the frequency distribution of sample averages is only one-tenth as large as the standard deviation of the individual IQ scores, it would equal one and one-half (or 1.5) points in this example.

We warned you that this was going to be a shaggy dog story. But don't give up yet. We're almost to the point, and once you see it, you'll understand a good bit of inferential statistics.

When a set of scores has a frequency distribution that follows a bell-shaped curve (the normal distribution), we know a lot about the scores. For example, a little more than two-thirds of the scores (68 percent to be exact) will be less than one standard deviation from their average value. In this example, then, we know that 68 percent of the sample averages will fall in an interval that is 1.5 points below to 1.5 points above the average of their frequency distribution. However, we've already said that the average of the sample averages will equal the average IQ score of the 5000 fourth-graders. Remember that this is the population parameter we're trying to estimate. So 68% of the sample averages will be within 1.5 points of the population average. Another useful fact is that 95 percent of the sample averages will fall in an interval that is between two standard deviations below the average of their frequency distribution and two standard deviations above that average. In this case, two standard deviations equals 3 IQ points, since two times 1.5 equals 3.0. So 95 percent of the 1000 sample averages will fall no farther than 3 points from the average IQ score of the population of 5000 students. Again, this fact is illustrated in Figure 3.

The only tricky part of understanding confidence intervals is coming next, so read the next section carefully.

Suppose that we consider any one of the sample averages that

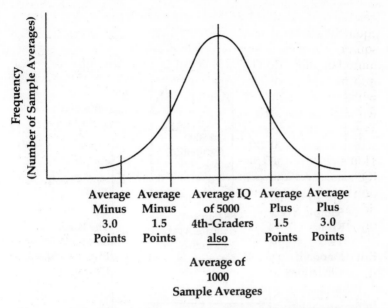

Figure 3. **Hypothetical Frequency Distribution of 1000 Sample Averages. Each Sample Consists of 100 Fourth-Grade Students, Selected From a Population of 5000 Fourth-Grade Students.**

is no more than two standard deviations from the true average IQ of the 5000 students. One such sample average is shown in Figure 4. If we were to form an interval by adding two standard deviations to that sample average, and then subtracting two standard deviations from that sample average, the interval would contain the true average IQ of the 5000 students. Look at Figure 4 to verify this fact. If we had chosen to start with any of the 95 percent of sample averages that are within two standard deviations of the average of the frequency distribution, we could have said the same thing. The interval we would have formed would contain the true average IQ test score of the 5000 students. Only if we chose one of the 5 percent of sample averages that was more than two standard deviations from the average of the frequency distribution, would our interval fail to contain the average IQ-test score of the population of 5000 fourth-graders. This is illustrated in Figure 5.

If you've followed the argument so far, you're about to under-

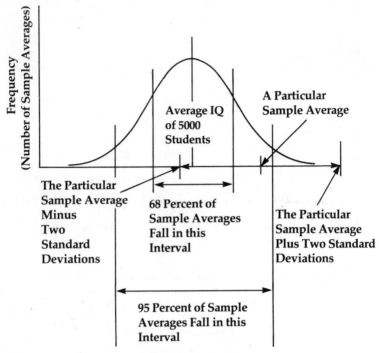

Figure 4. Hypothetical Frequency Distribution of 1000 Sample Average IQ Scores. A 95 Percent Confidence Interval Around a Particular Sample Average Includes the True Average IQ of the Population of 5000 Fourth-grade Students.

stand confidence intervals. Let's go back to reality. In fact, we're only going to select one sample of 100 students, not 1000 samples. For that one sample, suppose we completed the following three steps: (1) We computed the sample average IQ score; (2) we added two standard deviations to that average (to compute an upper confidence limit); (3) we subtracted two standard deviations from that average (to compute a lower confidence limit). We could then say that we were 95 percent confident that the population parameter—the true average IQ of all 5000 fourth-graders in the school system—fell between the lower confidence limit and the upper confidence limit. Now we wouldn't know for sure that we were making a true statement. We would only know that, had

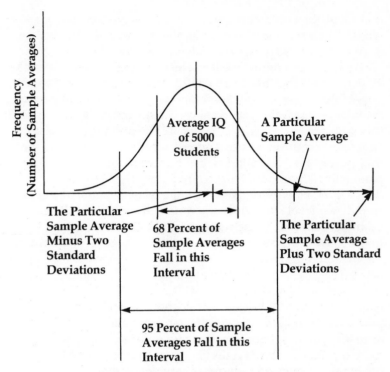

Figure 5. Hypothetical Frequency Distribution of 1000 Sample Average IQ Scores. A 95 Percent Confidence Interval Around Particular Sample Average Fails to Include the True Average IQ of the Population of 5000 Fourth-grade Students.

we selected 1000 independent samples of 100 students and followed the same procedure with each one, the true average IQ score for the 5000 fourth-graders would have fallen between the upper and lower confidence limits for 95 percent of those samples (that is, 950 out of 1000). The true average IQ score for the 5000 students would not have fallen between the upper and lower confidence limits for 5 percent of the samples (that is, 50 out of 1000). We have no way of knowing whether the one sample we selected was one of the "good" 950 for which the population parameter fell inside the confidence interval, or was one of the "bad" 50 samples for which the true average IQ score of the 5000 students was outside the confidence interval. Since our statement that the confidence interval contains the true average IQ would be correct for 95 percent of the intervals, we say that we have formed a "95 percent confidence interval."

The 95 percent figure is called a **level of confidence.** Although 95 percent is often used as a confidence level, it is not the only possibility. In some evaluation reports, you will find 90 percent confidence intervals, 99 percent confidence intervals, and sometimes, 80 percent confidence intervals. The confidence level used should reflect the evaluator's or researcher's desire for assurance that he or she has included the true value of the population parameter. The more assurance required, the higher the confidence level. However, given the same set of data, the higher the confidence level, the wider the confidence interval will be. For example, based on a single sample of 100 students, we might say that we are 95 percent confident that the average IQ of the entire population of 5000 students falls between 102.0 and 108.0. If we wanted a 99 percent confidence level and had the same sample of data, our confidence interval would range from 101.13 to 108.87. Notice that the 99 percent confidence interval is almost two points wider than the 95 percent confidence interval. In statistics as in life, you don't get something for nothing. If you want to be more confident without collecting more data, you have to widen the limits of your uncertainty (the confidence interval).

Interpreting Confidence Intervals in Evaluation Reports

The logic we've just discussed is applicable to all interpretations of confidence intervals, regardless of the context in which they're encountered. We've discussed the interpretation of a confidence interval around a single sample average. That's just one of many possible applications. Confidence intervals can be constructed for the purpose of estimating any number of population parameters, such as correlation coefficients, standard deviations, variances, and medians, to name just a few. One frequent application is a confidence interval on the difference between two population averages. Let's explore that application.

Suppose that you were interested in the relative effectiveness of two reading curricula you were considering adopting for third-graders in your school system. Since you'd been using the California Achievement Test (CAT) to assess third-grade reading achievement for the past several years, you would like to select the curriculum that would maximize your third-graders' average score on the CAT. Suppose that you select 100 students at random from the entire population of third-graders in your system, and then randomly assign 50 third-graders to each of the two competing curricula for an entire school year. At the end of the year, you have the CAT administered to all 100 students, and find that the 50 students who completed Curriculum A had an average reading comprehension score that was five points higher than the average of those who completed Curriculum B. For the 100 third-graders in your experiment, you are sure that Curriculum A was more effective, on average, than was Curriculum B. But that's not what you're interested in. You'd like to have some reasonable assurance that Curriculum A would be better than Curriculum B (i.e., it would result in a higher average reading comprehension score) if you used it with the entire population of third-graders in your school system. You can get the assurance you want, based on the data collected in your experiment, by computing a confidence interval.

It is possible to compute a confidence interval on the average reading comprehension score of all third-graders using Curriculum A, minus the average reading comprehension score of all third-graders using Curriculum B. The details of the computational procedures are not important here; what matters is that the confidence interval can be computed. Suppose that, when a 95 percent confidence interval was computed, the lower confidence limit was 2 points and the upper confidence limit was 8 points. You could then be 95 percent confident that Curriculum A would yield an average reading comprehension score for your entire population of third-graders that was at least two points higher than the average that would have resulted, had you used Curriculum B. In fact, the advantage in average score provided by Curriculum A might be as much as eight points, based on the data resulting from your experiment and a 95 percent confidence level.

Can you be sure that Curriculum A would be better, were you to use it with all of your third-graders? No. The level of confidence is 95 percent, not 100 percent. The only way to know for sure that one treatment is better than another would be to try both with the entire population of interest. Whenever you use statistical inference, some degree of uncertainty remains. That is the price you pay for collecting data from a relatively small sample, instead of the entire population. Can you say that the chances are 95 percent that Curriculum A would give you an average reading score for the population that was between two points and eight points higher than the average reading score you'd get using Curriculum B? Again, the answer must be no. Were you able to use Curriculum A with all of your third-graders, then wipe out their memories and start fresh using Curriculum B, there probably would be some real difference between their average scores on the CAT Reading Comprehension Test. That difference might favor Curriculum A by two points to eight points, or it might not. You have no way of knowing for sure. If it does, your confidence interval is correct—that is, the difference between the averages for Curricula A and B would be between two points and eight points. Otherwise, your confidence interval is not correct. Recall that there is no way of knowing whether a particular 95 percent confidence interval provides a true statement or a false statement. All you know for sure is the following: Had you repeated the experiment over and over thousands of times, computing a confidence interval each time, the true difference between the average achievement resulting from Curriculum A and that resulting from Curriculum B would have fallen within 95 percent of the confidence intervals you computed.

When you see a confidence interval in an evaluation report or a research report, then, interpret it the same way. Realize that the true value of the population parameter under consideration might or might not fall within the reported confidence interval. You can be sure, however, that a percentage of all possible confidence intervals, equal to the level of confidence, would contain the true value of the population parameter.

Confidence intervals have an added advantage. They provide information on the range of a researcher's or evaluator's uncertainty. In the example we just discussed, the confidence interval on the difference between reading comprehension averages had a lower limit of two points and an upper limit of eight points. Thus the experimental results leave you with a six point range of uncertainty, at the 95 percent confidence level. Now it could be the case that a two-point difference in average scores is trivial, as far as you're concerned, wherease, an eight-point difference is worth looking into. If that were the case, your experiment would have told you that Curriculum A was probably no worse than Curriculum B in terms of average reading comprehension score, but the difference in effectiveness might be trivial or it might be substantial. You can't know for sure on the basis of an experiment involving 100 students. But you can conclude that Curriculum A is a better bet than Curriculum B.

A Short Review Quiz

If you've survived this far, you've encountered a load of new vocabulary and some of the most challenging concepts in inferential statistics. It's probably a good time for a bit of review and the opportunity to test your grasp of the new material. In the following paragraphs, you'll find some hypothetical statistical information of the sort that you're likely to encounter in evaluation or research reports. Read each paragraph carefully, and write down your answers to the questions that follow. All of the answers are given at the end of the last set of questions. We know that, as a mature administrator who sets an example for those you supervise, you won't even be tempted to look at the answers in the book before you write down your own. Did we even suggest that wouldn't be fair?

Example A.
The Effectiveness of an Inservice
Science Education Program

In 1978 the Pomegranite, Ohio, Union School District conducted a series of inservice science education workshops for elementary school teachers to increase their awareness of curriculum materials on new developments in the physical sciences. The workshops had the parallel objective of increasing the teachers' knowledge of recent developments in the sciences that have immediate application to everyday life, such as medical applications of biotechnology (recombinant DNA applications, gene splitting, etc.) and computer applications of physics and chemistry (work with integrated circuits, transistor technology, etc.)

The success of the workshops in increasing teachers' knowledge of recent scientific developments was evaluated in part by administering a specially developed science awareness test to all participating teachers at the beginning of the first workshop, and then administering a parallel form of the test following completion of the last workshop. The evaluator reported the mean and standard deviation of teachers' scores on the pretest and the posttest, and then computed a 95 percent confidence interval on the teachers' mean "gain" from pretest to posttest. This approach to evaluation is not at all unusual, despite the fact that it is a very weak design. (But that's a topic for later discussion.)

The data reported by the Pomegranite Science Workshop evaluator are as follows. Review the data and then answer the questions that follow the table.

Table A. Results of Science Knowledge Evaluation for 30 Teachers Participating in Pomegranite Science Program Inservice Workshops.

PRETEST (25 items)	POSTTEST (25 items)	GAIN
Average = 17.5	Average = 23.5	Average = 6.0
Standard Deviation = 4.3	Standard Deviation = 5.0	
95 Percent Confidence Interval on Average Gain: (2.5, 9.5)		

Questions for Example A.

1. What group constitutes the sample in this example?

2. Is it clear that the sample was chosen randomly from some population?

3. To what population might the 95 percent confidence interval refer?

4. What is the population parameter that is of central interest in this example?

5. How confident can you be that the average gain of the population falls between 2.5 points and 9.5 points?

6. If you conducted the Pomegranite inservice evaluation independently, 1000 times, how many of the 1000 95-percent confi-

dence intervals would not include the average science achievement gain for the population?

Example B.
Estimating the Budget for the Oak Park School District—The Need for Remedial Classes.

At its last meeting, the Oak Park School Board recommended that any third through sixth-grade student whose arithmetic achievement score fell below the 35th percentile on the national norms of the Metropolitan Achievement Test be considered for placement in a remedial mathematics class. To estimate the need for remedial mathematics classes, the Research Department reviewed the most recent test data of 100 randomly selected students in each of grades three through six. The proportions of students in these samples whose arithmetic achievement fell below the 35th percentile were calculated, and then multiplied by the total numbers of students in these grades, to estimate the numbers of students who, by the school board's criterion, needed remedial mathematics. These estimated numbers of students needing remediation were then divided by 25 (the average class size in the Oak Park District) to estimate the need for remedial math classes by grade. Ninety percent confidence intervals on the number of classes needed in each grade were then computed. Results are shown in the following table.

Grade	Sample Size	Number <35%ile	Percent <35%ile	Estimated Number Needed	Estimated Classes Needed	LCL Classes	UCL Classes
3	100	20	20	1000	40	32	48
4	100	25	25	1200	48	42	54
5	100	28	28	1200	48	42	54
6	100	26	26	975	39	32	46

Questions for Example B.

1. What are the populations in this example?
2. What are the population parameters in this example?

3. What is the point estimate of the number of fifth-graders who need remedial mathematics classes?

4. What is the point estimate of the number of remedial math classes needed for fourth-grade students?

5. What is the lower 90 percent confidence limit on the number of remedial math classes needed for third-graders?

6. What is the upper 90 percent confidence limit on the number of remedial math classes needed for fifth-graders?

7. If you were trying to get the school board to approve a budget you were reasonably certain would be adequate, how many sixth-grade remedial math classes would you have them include?

Answers

Example A.

1. The 30 participating teachers constitute the sample in this example.

2. It is not clear that the sample was chosen randomly from some well-defined population. We can assume that it was not. You should realize that this situation is not at all unusual in research and evaluation, and that you'll encounter it in many evaluation reports. Here's how you handle interpretation. You are being asked by the author to assume that the sample he or she used is representative of some (unspecified) population. You must assume that the population is much like the sample in terms of its composition. For example, if you were to assume that number of years teaching were relevant to teachers' development of science awareness, then you would try to generalize the results shown for the sample to a population that was similar to the sample, in its distribution of years of teaching experience.

3. The population in this example is composed of all teachers who are potential participants in the science education inservice workshops.

4. The population parameter that is of central interest in this example is the average gain on the science awareness test that would be exhibited by the population of potential workshop participants. Your particular interest might be in the average

science awareness gain your population of teachers would experience, were you to use the inservice workshops in your school system.

5. You would have 95 percent confidence, since that is the level of confidence associated with the reported confidence interval.

6. 50 of the confidence intervals would fail to include the average gain of the population, since 50 is five percent of 1000.

Example B.

1. The relevant populations consist of all third-graders, fourth-graders, fifth-graders and sixth-graders enrolled in the Oak Park School District.

2. The population parameters of interest are the numbers of remedial mathematics classes needed by Oak Park's third-graders, fourth-graders, fifth-graders and sixth-graders.

3. The point estimate of the number of fifth-graders who need remedial mathematics instruction is 1200.

4. The point estimate of the number of remedial math classes needed for fourth-graders is 48.

5. The lower 90 percent confidence limit on the number of remedial math classes needed for third-graders is 32.

6. The upper 90 percent confidence limit on the number of remedial math classes needed for fifth-graders is 54.

7. You should have the school board budget for 46 remedial math classes for sixth-graders, since 46 is the upper 90 percent confidence limit on the number that will be needed. In that way, you can be reasonably sure that this aspect of the budget will be adequate, while at the same time, not going too far overboard. In fact, you could readily justify your budget figure here, based on a sound statistical analysis. The school board would surely resonate to your scholarship and logic.

Compare your answers with those given here. If you answered at least 10 of the 13 questions correctly, you can probably assume that you have an adequate grasp of the fundamental vocabulary of statistical inference, and a fair understanding of the logic of point and interval estimation. Those accomplishments are reserved to relatively few individuals in the world, so you can be proud of yourself and move ahead to the task of understanding the logic of hypothesis testing. If you missed more than three questions, don't give up! The subject matter is not trivial. Go back to the beginning of the discussion on estimation and review. Or review the vocabulary, if you feel it would be helpful. Then re-read the examples and the questions that follow them, and reconsider your original answers in light of the ones we have provided. Then, go on to the next chapter and learn all about hypothesis testing.

Problems—Chapter 7

7.1 In a report on sources of revenue in medium-sized school systems throughout the nation, Spend and Squander (1968) estimated the proportion of funds available from such sources as local tax revenues, state foundation programs, and federal programs. They based their estimates on a survey of superintendents of 100 medium-sized school systems, randomly selected from the entire U.S. population of such systems. One kind of information provided in the report was a 95 percent confidence interval on the percentage of school system funds provided by state foundation programs. The interval was reported as ranging from 43.2 percent to 67.8 percent.

Which of the following is a correct interpretation of this information?

    a. If a medium-sized school system is randomly selected from the population of all such systems in the United States, the probability is 0.95 that the percentage of its funding provided by a state foundation program will be between 43.2 percent and 67.8 percent.

    b. The percentage of funding in all medium-sized school systems in the U.S. that is obtained from state foundation programs might or might not be between 43.2 and 67.8, but one can be 95 percent confident that this interval includes the true value.

    c. Five percent of the medium-sized school systems in the nation obtain more than 67.8 percent of their funding from state foundation programs.

    d. Ninety-five percent of the school systems in the nation obtain at least 43.2 percent of their funding, but no more than 67.8 percent of their funding, from state foundation programs.

7.2 Administrators in a school system wanted to know the average length of time microcomputers would last (until they needed major repairs) if the computers were purchased for use in classrooms. To obtain this information, the administrators leased 20 microcomputers from a major manufacturer, had them installed in a random sample of classrooms throughout the school system, and recorded the number of days of operation until each microcomputer suffered a major breakdown. The average turned out to be 19.4 days.

Which of the following provides the best description of this result?

   a. The value 19.4 is a point estimate of the average time until major repairs would be needed, were micro-computers to be installed in classrooms throughout the school system.
   b. The value 19.4 is a population parameter. It is the expected number of days until failure that would be experienced, if the school system were to install microcomputers in all of its schools.
   c. The value 19.4 is a lower confidence limit on the number of days until failure that would be experienced, were the school system to install microcomputers in all of its classrooms.
   d. The value 19.4 is nothing more than a sample statistic. It has nothing to do with the average number of days to failure that would be experienced, were the school system to install microcomputers in all of its classrooms.

7.3 Meteorologists always seek new variables that help predict weather phenomena. In one recent study, researchers quantified sunspot activity, on a scale of 1 to 10, and correlated this sunspot rating with total rainfall, during each week of a year. The correlation was +0.42, across all of the weeks in a year.

   a. Do you think that this study is an example of descriptive or inferential research? Justify your answer.
   b. If you were to consider this an inferential study, what would the population be?
   c. If you were to consider this an inferential study, what would the sample be?

d. Would you describe the correlation of +0.42 to be a sample statistic or a population parameter?

7.4 Those who engage in games of chance involving sports usually inspect data on past performance very carefully. For example, in the not-so-distant past, the Dodgers and the Yankees have opposed each other in 28 baseball games. Of these contests, the Yankees won 21; 75 percent of the games. In predicting the outcome of future encounters, a sports enthusiast might use these data to compute a 95 percent confidence interval on the proportion of games the Yankees will win in an entire population of games that includes the future, as well as the past. Suppose that the 95 percent confidence interval ranges from 52 percent to 98 percent.

Which of the following conclusions is warranted on the basis of these data?

a. The Dodgers will never win another game when they meet the Yankees in the future, since the lower confidence limit is greater than 50 percent.

b. The probability that the Yankees win the game, the next time they meet the Dodgers, is at least 0.52.

c. Since the upper confidence limit is 98 percent, a value that is less than 100 percent, the Dodgers have some small chance of winning future games.

d. A person who engages in games of chance would, on the basis of these data, be better advised to bet on the Yankees than the Dodgers, since the lower confidence interval on the proportion of past and future games won by the Yankees is above 50 percent.

8.

The Logic of Hypothesis Testing

The Lay of the Land

If you understood the logic presented in the chapter on estimation, you've gone a long way toward an understanding of hypothesis testing as well. Not that the two forms of inference are identical. Indeed, they seek answers to fundamentally different questions. However, they both make use of concepts like statistics, parameters, samples, and populations. And distributions of sample statistics (such as the normal distribution of sample averages you encountered in the last chapter) play a crucial role in both procedures.

In estimation, the fundamental question we try to answer is, "What is the value of a population parameter?" In hypothesis testing, the fundamental question is, "Is it reasonable to believe that the value of a population parameter is x?" The "x" in that question is just a number, like zero, or 100, or some other value that makes sense in the context of a particular problem.

Hypothesis testing might be easier to understand if we consider a specific example. You're probably tired of it by now, but the example involving the average IQ-test score of all fourth-graders in your school system is as good as any, and at least it's familiar.

Let's suppose that you have an extensive achievement testing program in your school system, but your use of IQ tests is limited. Suppose that the average reading achievement of your fourth-graders has fallen far below that of previous years, and that you're searching for an explanation. You call a meeting of all fourth-grade teachers to explore the problem, only to have the teachers assert that your district is faced with an unusually dull crop of fourth-graders this year. The teachers suggest that the average verbal IQ of this year's fourth-graders is far different from the national average, and that's why reading achievement has suffered.

Your reaction to this distressing news is skepticism, since distributions of IQ-test scores don't change much from one year to the next unless a school system is affected by changes in its attendance area or substantial migration of families. Since your school system's boundaries and population have remained fairly stable, you decide to have your research department check out the fourth-grade teachers' claim.

Your testing budget is limited, as always, so you authorize testing a sample of 400 fourth-graders rather than all 5000 of them. To keep an open mind, you suggest that the research department begin with the assumption that your district's fourth-graders have an average verbal IQ that is exactly the same as the average for fourth-graders nationwide (100 points), and seek any evidence that refutes the assumption.

Regardless of whether you knew it at the time, you've called for a test of the **hypothesis** that the average verbal IQ of your district's fourth-graders is 100 points. The logic of your assignment to your research department is an exact parallel to that of **hypothesis testing.** You've begun with an initial statement of belief: an assertion that the average verbal IQ of your district's fourth-graders equals 100. Were the statement true, the fourth-graders' average IQ would be just the same as that of fourth-graders throughout the nation. A statement of belief such as yours is known in statistical language as a **null hypothesis.** It is a statement about the value of a population parameter (in this case, the value of the average verbal IQ of the population of fourth-graders in your school district). In hypothesis testing, belief in the validity of the null hypothesis continues, unless evidence collected from a sample is sufficient to make continued belief appear unreasonable. You've instructed your research department to seek just such evidence.

You should realize that the population parameter involved in your hypothesis test has some true value, although you cannot know this value unless you test every fourth-grader in your school district. Now the average verbal IQ of all 5000 fourth-graders is either equal to 100, in which case your null hypothesis is true, or it equals some other value, in which case your null hypothesis is false. If your null hypothesis is false, an **alternative hypothesis,** in this case, that the average verbal IQ of your district's 5000 fourth-graders does not equal 100, must be true. There is a law in logic, called the "law of the excluded middle," that asserts that a thing must either be true or false. You can't have it both ways. Now that might appear obvious, but it's important to the logic of hypothesis testing. Remember, at the start of this example your fourth-grade teachers asserted that the average verbal IQ of your fourth-graders was not equal to 100. Your response was to instruct your research department to begin with the converse of the teachers' assertion—that is, a belief that fourth-graders' average verbal IQ equalled 100—and to seek evidence refuting that belief. In short, you asked your research department to attempt to "nullify" your initial statement of belief, in which case the teachers' assertion, that the fourth-graders' average verbal IQ does not equal 100, must be true. The law of the excluded middle guarantees it.

So far, we've illustrated two basic concepts in hypothesis testing—a null hypothesis and an alternative hypothesis—and the logic that connects them. The logic may seem indirect and cumbersome. If you want to examine a claim that fourth-graders' average verbal IQ is different from the national average, why not do it directly, instead of stating just the opposite (the null hypothesis), and then trying to show that the null hypothesis is unreasonable? There are several answers to this question. The best answer is that you can't "prove" anything using inferential statistics, least of all that a population average is different from 100. Furthermore, a wealth of statistical theory, coupled with some reasonable assumptions, allows you to examine the reasonableness of your null hypothesis, and thereby determine whether the alternative is a better bet. In short, we don't know how to go at the problem directly, and we do know how to tackle it indirectly. Although the logic may appear to be convoluted and cumbersome, it's sound.

The General Procedure

Like estimation, hypothesis testing is an inferential procedure. To test a hypothesis, we use observations or measurements collected on a sample to decide whether or not our hypothesis about the value of a population parameter is reasonable. We decide either to continue believing that our null hypothesis is true, or we reject that belief, and decide that the alternative hypothesis is true. The steps of the process are as follows:

1. Identify the population and the population parameter of interest.

2. Define a null hypothesis and an alternative hypothesis.

3. Collect data on a random sample of objects or individuals, selected from the population of interest.

4. Compute a sample statistic that is an estimate of the population parameter of interest.

5. Depending on the value of the sample statistic, decide that the null hypothesis is true, or reject the null hypothesis in favor of the alternative hypothesis.

Boiled down to its essentials, that's the process of hypothesis testing. As you might suspect, we've oversimplified it just a tad. There are a few important questions that we've conveniently ignored, such as: What happens if you make the wrong decision? How do you determine exactly when to reject the null hypothesis, and when to continue believing it's true? How large a sample should you take? We know you can hardly wait to have all of these questions answered, so read on! Your fondest desires will be realized!

Errors and Risks

The outcome of hypothesis testing is a decision; in the case of this example, a decision to continue believing that your district's fourth-graders have an average verbal IQ of 100 (the null hypothesis), or a decision to reject that belief, in favor of the alternative hypothesis that the fourth-graders' average IQ differs from 100. In general, the outcome of hypothesis testing is a decision to continue believing the null hypothesis, or a decision to reject the null hypothesis, in favor of a belief in the alternative hypothesis.

Reality exists, regardless of your decision. That is, the population average is either equal to 100 (in which case the null hypothesis is true) or it is equal to some value other than 100 (in which case the alternative hypothesis is true). Since the decision you will make in this example will be based on the test performance of only 400 fourth-graders, and not on the performance of the entire population of 5000, you might make the wrong decision. That is, you might decide that the average IQ of all 5000 fourth-graders is different from 100, when in fact, it equals 100. Such a mistake has a name in hypothesis testing. It is called a **Type I error.** A Type I error occurs when the null hypothesis is rejected even though it is true.

One of the joys of hypothesis testing is that the risk of making a Type I error can be set by the researcher or evaluator. You can design and adopt a decision rule that ensures that you will incur a very small risk of making a Type I error. The risk of making a Type I error is called the **level of significance** of the hypothesis test. The symbol that is used for the level of significance is a lower-case Greek letter alpha (α), so the level of significance is sometimes called the **alpha level** of the test. When you read an evaluation report or a research report and see that a particular statistic is "significant at the .05 level," the author is telling you that a null hypothesis was tested in such a way that the risk of making a Type I error was set at .05 (five percent), and the null hypothesis was rejected in favor of the alternative hypothesis. If this last bit of information seemed a bit opaque, don't be too concerned. There will be many more illustrations of hypothesis testing and interpretations of "significant" statistics in later sections of the book.

For the moment, let's get back to errors and risks. So far we've described one kind of error (Type I), the error that occurs when the decision-maker decides that the null hypothesis is false (e.g., that the average verbal IQ of the district's fourth-graders is not equal to 100), when in fact it is true. This is the only type of error that can be made if the null hypothesis is rejected. Suppose, however, that the decision-maker decides that the null hypothesis is true. In doing so, she or he will be rejecting the alternative hypothesis. In any given circumstance, that decision can be appropriate, or it can result in another type of error. If the alternative hypothesis is true, but the decision-maker decides to stick with the null hypothesis, a **Type II error** results. In our present example, suppose that, on the basis of advice from the research department, the administrator decides that the average verbal IQ of his or her district's fourth-graders is equal to 100 when, in fact, the population average is different from 100. The administrator would have committed a Type II error. The risk of making a Type II error doesn't have a special name in statistics, but the Greek letter beta (β) is usually used to denote it.

Once again we've discussed a number of complex concepts that might not seem to hang together in a logical pattern. You may well be asking yourself: "Where are they going with all those hypotheses, errors and risks?" Have faith! Remember that the logic of confidence intervals turned into a shaggy dog story, but we did get to the point eventually. The same thing will happen with hypothesis testing. If you stick with it, you'll finally be able to figure out all of those "*" and "**" marks next to the numbers in evaluation and research reports, and you'll understand the cryptic footnotes that say something like "$p < .05$" or "significant, alpha $= .01$." There now, doesn't that inspire you?

Let's review. In any hypothesis testing problem there are two hypotheses to deal with. The *null hypothesis* is an initial statement of belief about the value of a population parameter. The object of hypothesis testing is to decide whether it is reasonable to continue believing that the null hypothesis is true, or to reject that belief, in favor of an *alternative hypothesis*. The decision is based on data collected from a sample that is randomly drawn from the population.

As in life, our decision can be either right or wrong. There are two ways in which we can be right, and two ways in which we can be wrong. If the null hypothesis is, in fact, true, we can be right by reaching that decision. We can be wrong by deciding that the null hypothesis is false, despite the fact that it is true. That kind of error is called a Type I error. The other possibilities occur if the null hypothesis is actually false and the alternative hypothesis is true. Here again, we can be right by reaching that decision. Or, we can be wrong by deciding that the alternative hypothesis is false, despite the fact that it is true. That kind of error is called a Type II error. These four possibilities are summarized in the table shown below.

Earlier, we related one of the primary joys of hypothesis testing. The researcher or evaluator can control the risk of making

THE POTENTIAL OUTCOMES OF A HYPOTHESIS TEST

The True State of Reality

		The Null Hypothesis is True	The Alternative Hypothesis is True
D E C I S I O N	The Null Hypothesis is True	Probability $=1-\alpha$	Type II Error (Risk β)
	The Alternative Hypothesis is True	Type I Error (Risk α)	Probability $=1-\beta$

errors. One of the steps in hypothesis testing is to develop a decision rule—a rule that specifies what values of a sample statistic will lead you to continue believing that the null hypothesis is true, and what values of a sample statistic will lead you to reject the null hypothesis. It is possible to design the decision rule so that the risk of making a Type I error is as small as you want. Furthermore, if your sample size is large enough, you can make the risk of making a Type II error small at the same time you make the risk of making a Type I error small. In other words, it is possible to design a hypothesis test so that you make the right decision most of the time.

Decision Rules

Let's return to our example. Your research department followed your instructions, and dutifully administered an IQ test to 400 fourth-graders who were randomly sampled from the 5000 in your school district. Suppose that the average IQ of the 400 sampled fourth-graders was 101.2. Now what should you do about your null hypothesis that the average IQ of the entire population of fourth-graders is 100? Well, 101.2 is pretty close to 100, isn't it? Don't you think it is quite possible to find a sample of 400 fourth-graders with an average IQ of 101.2 in a population of

5000 fourth-graders who have an average IQ of 100? Of course it is! So based on the data collected by your research department, you really don't want to reject the null hypothesis that fourth-graders in your school district have the same average IQ as fourth-graders throughout the nation.

Now suppose that the IQ test results for the sample had been somewhat different. Let's say that the average IQ for the 400 sampled fourth-graders turned out to be 85.6. That's a long way from 100. Do you think that it is possible to find a sample of 400 fourth-graders with an average IQ of only 85.6 in a population of 5000 fourth-graders who have an average IQ of 100? Well, it's possible, but it's not at all likely. In fact it's so unlikely that the only sensible thing to do is reject your initial belief that the population of fourth-graders have an average IQ of 100. In other words, you ought to reject the null hypothesis, and conclude that your fourth-grade teachers are right. You do have a dull crop of fourth-graders this year!

So far we've discussed two possible outcomes of giving a sample of 400 of your fourth-graders an IQ test. In the first case, the average IQ of the sample was very close to the value you'd hypothesized for the population. The only rational behavior was to persist in your belief that the average IQ of the population was indeed equal to 100 (i.e., continue to believe that the null hypothesis was true). In the second case, the average IQ of the sampled fourth-graders was substantially different from the value you'd hypothesized for the population. The only rational behavior was thus to reject your belief that the population had an average IQ of 100 (i.e., reject the null hypothesis). Both of these cases are so extreme that the appropriate decision was obvious. But what decision would you reach if the average IQ of the 400 sampled students was neither very close to, nor very far from, the hypothesized value of 100? For example, what if the sample average IQ turned out to be 103? Should you reject the null hypothesis or retain it? If you're like most people, your gut feelings aren't very strong either way. In fact, what appears to be a borderline decision might induce a bit of indigestion if you rely solely on your tummy to tell you what to do! What's needed is some reliable, scientific way of determining when you ought to retain the null hypothesis and when you ought to reject it. In short, you need a firm decision rule. What we're talking about is the kind of rule that says something like, "If the average IQ of the 400 sampled students is between 98.5 and 101.5, continue to believe that the null hypothesis is true; if the sample average IQ is either smaller than 98.5 or larger than 101.5, reject the null hypothesis." (At this point you're probably wondering where in the world those numbers came from. Hang in there, and we'll let you in on the secret.) Once you have a decision rule, you won't have to rely on your gut feelings, your antacid bills will go down, and you'll be able to justify your decision on scientific grounds.

Sampling Distributions

Now comes the story on how we came up with the decision rule we so glibly spouted in the last paragraph. As Johnny Carson often says, there's good news and bad news. The bad news is that it's a long story. The good news is that you've heard most of it before—back in the chapter on the logic of estimation.

Remember the hypothetical example in which we sampled over and over again from the same population of fourth-graders, each time computing the average IQ of the sampled students? Let's suppose that we were to do the same thing here. In the earlier example we selected samples of 100 students; in this example, to avoid boring you with total duplication, we'll select samples of 400 students. As before, we'll select a sample of students using simple random sampling; administer an IQ test to the sampled students; record their average test score; and return the students' names to the population prior to selecting the next sample. Actually, we're going to draw a large number of independent samples (1000 samples is as good an illustration as any), and record the average IQ for each sample. When we finish, we'll have a list of 1000 sample averages.

Back in the chapter on estimation, you learned that the frequency distribution of the 1000 sample averages would look like a bell-shaped curve called the normal distribution. That's still true in this chapter. Also, it is still true that the average of the 1000 sample averages would be precisely equal to the average IQ of the 5000 fourth-graders in the population (the unknown population parameter). Let's assume, as before, that the standard deviation of the IQ scores of the 5000 fourth-graders is equal to 15 points. Then the standard deviation of the frequency distribution of sample averages (the standard error of the mean) would be equal to three-fourths of a point (0.75). That number didn't pop out of the air by magic. It was found by dividing the standard deviation of the frequency distribution of IQ scores for individual fourth-graders (15 points) by the square root (20) of the sample size (400). The square root of 400 is 20, and 15/20 is three-fourths or 0.75.

If, for the moment, we assume that the null hypothesis is true—namely that the population of 5000 fourth-graders has an average IQ of 100—we can draw a picture of what the frequency distribution of sample averages would look like. Take a look at Figure 6, and try to recall what you learned about normal distributions in the chapter on estimation. First, we said that 68 percent of the 1000 samples will have averages that fall within one standard deviation of the average of the population. If the null hypothesis is true then, we can conclude that 68 percent of the samples will have averages that are between 99.25 and 100.75. We found these values by adding 0.75 to 100, and by subtracting 0.75 from 100. It is also true that 95 percent of the samples will have averages that are between 98.5 and 101.5, because 95 percent of a normal distribution falls within two standard deviations of the population average. If the standard deviation of the distribution is 0.75 points, the population average plus two standard deviations equals 100 + 1.5 = 101.5. Also, the population average minus two standard deviations equals 100 − 1.5 = 98.5.

The Risk of Making a Type I Error

We said earlier that you could set the risk of making a Type I error (the risk of rejecting the null hypothesis when it is true). In this example, you're really deciding the chance you want to take of concluding that the 5000 fourth-graders have an average IQ that is different from 100 when, in fact, their true average IQ equals 100. Or to put it another way, how big a risk do you want to take of concluding that the fourth-grade teachers were right, when in reality they were wrong? The answer to this question is called the "level of significance" of your hypothesis test.

A decision on an appropriate level of significance should take into account the seriousness of making a Type I error. The more serious the consequences of rejecting the null hypothesis when it is true, the smaller the risk you want to take of making a Type I

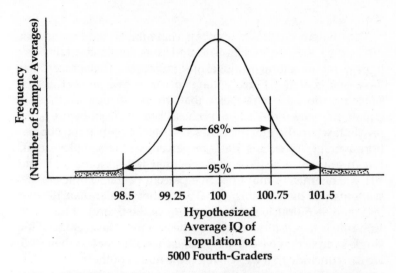

Figure 6. Hypothetical Frequency Distribution of 1000 Sample Averages, Assuming That the Null Hypothesis (Population Average=100) is True. Each Sample Consists of 400 Fourth-Grade Students, Selected from a Population of 5000 Fourth-Grade Students.

error. In many research and evaluation studies, a level of significance equal to five percent (0.05) is used. This is based partly on tradition, and partly on thoughtful consideration. It says that the researcher or evaluator is willing to reject a true null hypothesis only five percent of the time. Or to put it another way, the researcher or evaluator is satisfied knowing that he or she will appropriately retain a true null hypothesis 95 percent of the time.

If we adopt a five percent level of significance in this example, we can proceed to develop a decision rule.

Computing a Decision Rule

If we adopt a level of significance equal to five percent (0.05), our decision rule should be the one stated in the illustration above; namely, reject the null hypothesis that the population average IQ equals 100, on the basis of data from a single sample of 400 students, only if the sample average IQ is either greater than

101.5 or less than 98.5.

Now where did those numbers come from? You can see the answer by looking at Figure 6 and considering the following logic. First, in setting our level of significance at five percent, we have stated that we are willing to run a five percent risk of rejecting the null hypothesis that the population average IQ equals 100, when the null hypothesis is true. From Figure 6, we see that, when the null hypothesis is true (i.e., when the population average IQ equals 100), 95 percent of the samples of 400 fourth-graders will have averages that fall between 98.5 and 101.5. This means that only five percent of the samples of 400 students will have averages that are either greater than 101.5 or less than 98.5 (in the shaded portions of the figure). If our decision rule is to reject the null hypothesis only if the average of the single sample that we draw is greater than 101.5 or less than 98.5, we can conclude that we will reject the null hypothesis, when it is true, only five percent of the time. This is exactly the outcome we sought. The decision rule we have stated conforms precisely to the level of significance we desired. You will find that decision rules in hypothesis testing are always a consequence of the desired level of significance (risk of making a Type I error), so the consistency we found in this example is no accident.

Another Example—And A Review

In the chapter on estimation we used an example that's characteristic of many decision situations in school systems and other organizations: you have to choose between two programs, projects, or products, either of which might be best. In our example in the last chapter, you were trying to decide which of two third-grade reading curricula you should adopt. We suggested that a confidence interval on the *difference* between the average California Achievement Test scores in reading be computed for those third-graders taught using Curriculum A and those taught using Curriculum B. We examined the confidence interval to determine whether those taught using Curriculum A would do better or worse on the California Achievement Test (CAT) than those taught using Curriculum B, and how large the average achievement difference was likely to be. We found that the confidence interval had a lower limit of two points, and an upper limit of eight points, in favor of Curriculum A. We concluded that Curriculum A was better than Curriculum B in terms of average score on the CAT, by an amount that might be trivial (two points) or might be substantial (eight points).

Hypothesis testing provides an alternative statistical approach to decision-making when you have to choose between two programs, projects, products, or curricula, and it's impractical to try both possibilities with the entire population of students you're interested in.

To use hypothesis testing with the problem of choosing between Curricula A and B, you'd begin by defining a null hypothesis. Since you don't know which of the two curricula would provide a higher average score on the California Achievement Test for your school system's third-graders, your null hypothesis should be that the two curricula are equally effective. That is, you'd hypothesize that the third-graders' average CAT score would be exactly the same, regardless of whether they were taught using Curriculum A or Curriculum B. To express this hypothesis statistically, you would hypothesize that the difference between the third-graders' average CAT score using Curriculum A and their average CAT score using Curriculum B was equal to zero.

The alternative hypothesis in this example is the precise opposite of the null hypothesis. If two things aren't equal, they must be unequal. So your alternative hypothesis is that the average CAT reading score for third-graders would differ, depending on whether you used Curriculum A or Curriculum B. Now it's important to recognize exactly what we mean by this alternative hypothesis. What we're saying is the following: Suppose we were to have each third-grader in your school system study reading using Curriculum A, and then complete the CAT reading test. If we then calculated the average CAT reading score of all third-graders in your system, what we'd have would be a population parameter (the population average CAT reading score for all third-graders, taught using Curriculum A). Next, let's suppose we erased the memories of all of the school system's third-graders, so they had no recollection of being taught reading using Curriculum A, and instead we taught the entire population using Curriculum B. If at that point we were to administer the CAT reading test and then compute the students' average score, we'd have another population parameter (the population average CAT reading score for third-graders, taught using Curriculum B). The alternative hypothesis just says that these two population parameters would be different from each other, whereas the null hypothesis says that they would be equal. In practice, were we to decide to reject the null hypothesis in favor of the alternative hypothesis, we would be deciding that one of the two curricula was more effective than the other, at least as far as scores on the CAT reading test were concerned.

When we considered forming a confidence interval on the difference between third-graders' average CAT reading scores in the last chapter, we conducted an experiment to collect the data we needed. Recall that we first selected a random sample of 100 third-graders from the entire system-wide population, and then we randomly divided the sample between Curriculum A and Curriculum B. That is, we assigned 50 third-graders to A and 50 to B, for an entire school year. At the end of the year, we administered the CAT reading test to both groups, and computed the average score of those instructed using each curriculum. We use those sample averages (and some other numbers we didn't describe explicitly) to construct a confidence interval. Instead of computing a confidence interval on the difference between population averages, we could have used the same experimental data to test the null hypothesis that the two curricula were equally effective against the alternative hypothesis that they differed in effectiveness. In fact, it's generally true that the data collection procedure is exactly the same, whether we're constructing confidence intervals or testing hypotheses. It's what we do with the data once we've collected them that differs.

So far we've identified a population and the parameters that are of interest to us, and we've defined a null hypothesis and an alternative hypothesis. We've also taken the next step in hypothesis testing, which is to specify a plan for collection of data. If you look back to the section headed "The General Procedure," you'll see that the last two steps in hypothesis testing are to compute a sample statistic that is an estimate of the population parameter of interest, and to reach a decision on whether or not the null hypothesis is true. Of course you know that we've left out a few important specifics in defining these "five easy steps to an ideal hypothesis test." We must specify the risk we're willing to take of making a Type I error (α), and then we've got to develop a decision rule. Deciding on an appropriate value for α is pretty easy, but coming up with a decision rule will take a good deal more thought and discussion.

In testing a hypothesis that concerns the difference between two population averages, the logical statistic to use would seem to be the difference between corresponding sample averages. So why don't we just compute the average CAT reading score of the 50 third-graders who were taught using Curriculum A, and the average CAT reading score of the 50 third-graders who were taught using Curriculum B, and then compute the difference between the sample averages? If the sample averages are very similar, the difference between them will be close to zero and it would seem reasonable to continue believing that the null hypothesis is true (that Curricula A and B are equally effective in terms of average CAT reading scores). On the other hand, if the 50 third-graders taught using Curriculum A had an average CAT reading score that was much higher or much lower than the average CAT reading score of the 50 students taught using Curriculum B, we would be justified in rejecting the null hypothesis in favor of the alternative (that the two curricula are not equally effective). As in the example we discussed earlier, we have the sticky problem of finding an explicit decision rule. How different must the two sample averages be, to cause us to reject the null hypothesis in favor of the alternative hypothesis? We know the answer depends on where we set our Type I error level (α). But it also depends on how much the difference between sample means would fluctuate, were we to conduct our experiment over and over.

The logic used in developing a decision rule for this example is virtually identical to that used in the last example. To avoid complicating the description, we're going to make some assumptions that are a bit unrealistic but not really troublesome. In particular, we will assume that we know the standard deviation of scores on the CAT reading test for the entire population of third-graders in your school system, and that this standard deviation would be exactly the same, whether the students were taught using Curriculum A or Curriculum B. If you want to dazzle your friends or your boss or your lover with a truly impressive word, tell them that you're going to assume you have "homoscedasticity." Now wait, you don't have to run to the doctor for an immediate shot in the behind. Homoscedasticity just means equal spread or equal variability. So our assumption of equal standard deviations on the CAT, regardless of the curriculum we use, is a homoscedasticity assumption.

In this experiment, it would seem reasonable to take a very small risk of deciding that one of the curricula was better than the other if, in fact, they were equally effective. That's suggesting that we want a very small risk of making a Type I error. So let's set alpha equal to 0.01. In doing so, we're saying that we'll only have a one percent chance of rejecting the null hypothesis if the curricula are, in fact, equally effective.

As in the last example, we have to engage in a purely hypothetical discussion at this point in order to develop a decision rule. You're also going to have to accept a few things on faith. Here goes. Let's suppose we were to repeat our experiment of sampling 100 third-graders from the population of 5000 over and over, each time dividing the sample randomly into two groups of 50. And with each repetition we'd teach one group using Curriculum A and the other using Curriculum B, and then administer the CAT reading test at the end of the school year. We'd compute the average CAT score for each group, and the difference between the averages. If we did this over and over, say 1000 times, we'd have a list of 1000 differences between average scores using Curriculum A and average scores using Curriculum B. This list of 1000 differences between sample averages would have a frequency distribution, just like any list of 1000 numbers. Fortunately, statistical theory tells us what the frequency distribution would look like, and we can use this information to develop a decision rule for accepting or rejecting the null hypothesis.

The frequency distribution of differences between sample averages would look like a normal distribution (our old friend the bell-shaped curve). If the null hypothesis were true, the average of this frequency distribution would equal zero. This just says that if the population average of third-grader's CAT reading scores was the same, regardless of whether Curriculum A or Curriculum B was used (in which case the difference between these population averages would be zero), the difference between sample averages would fluctuate around an average value that was also equal to zero.

So far we know that the frequency distribution of the difference between sample averages would look like a normal distribution, and that its average would be zero, provided the null hypothesis was true. To construct a decision rule we need to know two additional facts: (1) the standard deviation of the frequency distribution, and (2) how many standard deviations from the average we'd have to go in order to include 99 percent of the distribution. (If we include 99 percent of the distribution, we'll only exclude one percent—an amount that is equal to our desired risk of making a Type I error). To answer the first question, we're going to use some invention and some statistical theory. Our final result will be a standard deviation that is equal to 1.6 points. The invention part comes from assuming that, if we were to give all third-graders in the school system the CAT reading test, the standard deviation of their individual test scores would equal 8 points. The statistical theory part comes from a formula that tells us how to calculate the standard deviation of the difference between sample averages, if we know the standard deviation of test scores for individual students. The formula isn't really important here, so we're not going to impress you with it. If you really want to see it, you'll find it on page 295 of Glass and Stanley (1970), and in lots of other introductory statistics books. The answer to the second question is 2.58. Ninety-nine percent of our 1000 repeated experiments would result in a difference between sample averages using Curricula A and B, that would fall within 2.58 standard deviations of the average of the frequency distribution. This fact comes from what's known about all normal distributions.

Here we are again, in the thicket of another shaggy dog story. Let's review the facts we know so far. The standard deviation of the frequency distribution equals 1.6 points. If we assume that the null hypothesis is true, the average of the frequency distribution will equal zero. And if we go 2.58 standard deviations on both sides of the average, we will include 99 percent of the frequency distribution. Now 2.58 standard deviations is just 2.58 times 1.6 points, or 4.13 points. Hang on! A decision rule is about to leap right out at us! To see it, look at Figure 7, which illustrates the facts we've just summarized.

Remember that we only wanted a one percent risk of rejecting the null hypothesis if Curricula A and B were, in fact, equally effective. So we wanted a decision rule that would cause us to retain the null hypothesis 99 percent of the time, if it was true. From Figure 7 we can see that our decision rule should be as follows: If the difference between sample averages is bigger than 4.13 points, either plus or minus, reject the null hypothesis. If the difference between sample averages is between −4.13 points and +4.13 points, continue believing that the null hypothesis is true. If we followed this decision rule, and if the null hypothesis were true, only one percent of the repeated experiments would result in a difference between sample averages that fell in the shaded portions of Figure 7, thus causing us to falsely reject the null hypothesis. Ninety-nine percent of the experiments would give us a difference between sample averages that was between −4.13 points and +4.13 points, causing us to retain a true null hypothesis, just as we should. So once again, our decision rule is totally consistent with our desired risk of making a Type I error. The values −4.13 and +4.13 are known in statistical parlance as **critical values,** and the shaded regions of Figure 7 are called **critical regions.** In general, if the sample statistic falls in a critical region, we reject the null hypothesis.

It is important to clearly separate the hypothetical data-collection scenario we've described in this example from actual practice in a real hypothesis-testing situation. Once again we've used the illustration of selecting 1000 independent samples of 100 third-graders just as a way of describing the frequency distribution of the difference between sample averages. In an actual hypothesis-testing situation, we'd only select one sample of 100 students, divide that sample in half, instruct one group of 50 students using Curriculum A, instruct the other group of 50 using Curriculum B, and collect achievement data from both groups. We would base our decision on whether to retain or reject the null hypothesis solely on the data collected from the one sample of 100 students, not on the data from 1000 samples. If the average CAT reading scores of the 50 students taught using Curriculum A differed by more than 4.13 points (in either direction) from the average CAT score of the 50 students taught using Curriculum B, we would reject the null hypothesis. If the two sample averages differed by less than 4.13 points, we would retain the null hypothesis. The discussion about 1000 samples of 100 students each was just a way of describing the basis of

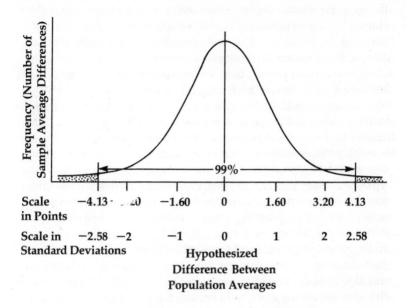

Figure 7. Hypothetical Frequency Distribution of 1000 Differences Be-
tween Sample Averages, Assuming That the Null Hypothesis
(Difference Between Population Averages=0) is True. Each
Sample Consists of 100 Third-Grade Students, Randomly
Split Between Two Curricula. Assumed Standard Deviation
of Individual Scores is Eight Points.

our decision rule.

Suppose that, when you conducted your experiment, you
found that the 50 students who were taught using Curriculum A
had an average CAT reading score that was 6.2 points higher than
the average CAT reading score of the 50 students taught using
Curriculum B. What would your decision be? Well, since 6.2 is
larger than 4.13, you should reject the null hypothesis that the
two curricula are equally effective. We would describe the 6.2-
point difference between the sample averages by saying that it is
"significant at the .01 level." This is just statistical shorthand for
saying that we have tested a null hypothesis using a Type I error
level of .01, and, on the basis of sample data, we have rejected the
null hypothesis. Notice that the difference in sample averages

(6.2 points) would fall in the shaded region of Figure 7 that is on the far right end of the curve. We've already said that we'll reject the null hypothesis if the difference between sample averages falls anywhere within either shaded area.

Suppose that we had been more tolerant of the possibility of making a Type I error and, instead of setting our alpha level at .01, we had set it at .05. This means that we would have been willing to run a five percent risk of making a Type I error. How would that have changed our decision rule? It's easy to determine our new decision rule because most of the computations wouldn't change at all. First, the frequency distribution of the difference between sample averages would be exactly the same. If the null hypothesis were true, the average of the frequency distribution would still equal zero. The standard deviation of the frequency distribution would still be 1.6 points. But instead of having the critical values 2.58 standard deviations from the average of the frequency distribution, they would only be two standard deviations from the average. Remember that 95 percent of the normal distribution falls between points that are two standard deviations below the average and two standard deviations above the average. By including 95 percent of the frequency distribution, we only exclude five percent, which is the new value we set for the risk of making a Type I error. In this example, since one standard deviation equals 1.6 points, two standard deviations would equal 3.2 points. So our new decision rule, with an alpha of .05, would be as follows: reject the null hypothesis only if the difference between sample averages is larger than 3.2 points or smaller than −3.2 points. In other words, if the average CAT reading score of either sample of 50 third-graders differed from the average of the other sample by more than 3.2 points, we would reject the null hypothesis. We would then say that the difference between sample averages was "significant at the .05 level." The critical regions, the areas under the curve in Figure 7 that should be shaded, would fall to the right of 3.2 points and to the left of −3.2 points.

It is worthwhile to compare the decision rule and the critical values when we set alpha at .01 with the decision rule and the critical values when alpha is set at .05. With an alpha of .05, we would reject the null hypothesis if the average CAT reading scores for the two samples differed by more than 3.2 points. With an alpha of .01, the two sample averages would have to differ by at least 4.13 points before we would reject the null hypothesis. This shows that the difference between sample averages could be significant at the .05 level (that is, greater than 3.2), but might not be significant at the .01 level (greater than 4.13). For example, if the average CAT reading score of those taught using Curriculum B was 35.0, and the average for those taught using Curriculum A was 31.5, the difference would be 3.5 points. Since 3.5 is greater than 3.2, the difference would be significant at the .05 level, and we would reject the null hypothesis that the two curricula were equally effective, at the five percent level of significance. However, 3.5 is less than 4.13, so the difference between sample averages would not be significant at the .01 level. In fact, if we were testing our null hypothesis at an alpha level of one percent, we would retain the null hypothesis. In all hypothesis testing, it is always the case that significance at the .05 level does not necessarily mean the statistic is significant at the .01 level. However, a statistic that is significant at the .01 level will always be significant at the .05 level. If you take another look at Figure 7, you can see why this is true. In order to be significant at the .01 level, a statistic must be farther from the average of the frequency distribution than it need be, if it is only significant at the .05 level. The further from the hypothesized value of a population parameter a statistic is, the smaller the alpha level at which it is significant.

Another Review Quiz

Well, it's that time again! Time to make sure that you have all those alphas, betas, hypotheses, risks, and errors straight in your mind, so you can really gobble research reports with confidence—authoritatively separating the wheat from the chaff (to put it most politely).

Once again, we'll provide a few hypothetical examples of research and evaluation reports that contain inferential statistical results of the form you'll encounter in the real world. There will be questions following each example. After reading each example carefully, write down your answers to the questions that follow. At the end of this thrilling exercise, you can compare your answers to those that follow the last example.

Example A.
The Validity of a Teacher Selection Test

In the 1980's world of public education, where accountability and economy are the watchwords, assignment of blame is a favorite sport. For example, in your school system the board has suggested that low basic skills achievement scores can be attributed directly to inadequate screening of teachers at the time they are hired. Their proposed solution is to institute a teacher competency test, to be administered to prospective teachers as a part of the interview process.

In response to the board, you've begun a search for an existing teacher competency test that might meet your school system's needs. A computerized search of Educational Research Information Center (ERIC) files yields an article on a test that was validated in a South Carolina school system. When you read the validation report, here's what you find:

The test was administered on an experimental basis to all applicants for teaching positions in a very large South Carolina school system, but applicants' scores were ignored when making hiring decisions because the validity of the test was unknown. At the end of one year on the job, the test scores of newly-hired teachers were compared with teaching effectiveness ratings given by their principals. The instrument for rating teaching effectiveness had been in use in the school system for a number of years. The researchers selected a sample of 25 teachers who had scored between zero and 50 on the competency test, and another independent sample of 25 teachers who had scored between 51 and 100 on the competency test. Statistics on principals' ratings of teaching effectiveness were computed and reported for both samples, and a hypothesis test was conducted to investigate the question of test validity. The data reported in the article were as shown in the table following the Questions for Example A.

Questions for Example A.

1. What populations are of interest in this example? To what populations are inferences being made?
2. What is the null hypothesis?
3. What is the alternative hypothesis in this example?
4. What is the risk of making a Type I error in this example?

5. Explain the statement that follows the asterisk at the bottom of the table: "significant at the .01 level."

6. Do the data in the table support the researcher's contention that the competency test is a valid measure of teaching effectiveness? Why or why not?

7. What kinds of inferences would you be making if you recommended use of the competency test in your school system on the basis of the reported validity analysis?

Group:	Scored Between Zero and 50 on Competency Test	Scored Between 51 and 100 on Competency Test
Average Effectiveness Rating by Principal:	64.8	82.6
Standard Deviation of Effectiveness Ratings by Principal:	8.2	7.8
Difference in Average Effectiveness Ratings:	17.8*	

*Significant at the .01 level.

Example B.
Teacher Salaries and Student Achievement — Are They Related?

In keeping with the old adage that you get what you pay for, an article in a recent issue of the *School Administrator's Quarterly* reported on a study of the relationship between elementary school teachers' salaries and the achievement of their students. A researcher in one of the largest school systems in California selected a random sample of 100 elementary school teachers from the list of 2571 teachers employed by the system, and looked up the sampled teachers' annual salaries. She then went to the system's achievement test files and recorded the average achievement of students in each teacher's class in four different subject areas: reading, spelling, arithmetic concepts, and arithmetic computation. The achievement scores reflected students' performances on system-developed criterion-referenced tests, and were scaled in terms of percent of items correct, so that scores would be compared across grade levels.

Wondering whether there was any correlation at all between teacher salary and student achievement, the researcher computed a series of correlation coefficients between teachers' annual salaries and the average achievement of their students, with the following reported results:

Correlation between annual teacher salary and:

Reading	.08
Spelling	.17*
Arithmetic Concepts	−.02
Arithmetic Computation	.20**

---

* significant, alpha = .10
** significant, alpha = .05

Questions for Example B.

1. Four separate null hypotheses were tested in this example. Can you identify them?

2. What were the alternative hypotheses associated with these null hypotheses?

3. To what population was the researcher attempting to generalize?

4. What was the population parameter in the hypothesis test concerning the relationship between teacher's salary and students' reading achievement?

5. Discuss the meaning of the statement, "significant, alpha = .05."

6. Does it appear that teacher salary and student achievement are related?

7. Do these results suggest that raising teachers' salaries is a good way to increase students' achievement? Why or why not?

Answers
Example A.

1. The populations that are of direct interest in this example consist of all first-year teachers in the South Carolina school system who scored between zero and 50 on the competency test, and all first-year teachers in the system who scored between 51 and 100 on the competency test. These are the populations from which the researcher selected his samples. Formally, inferences always refer to the populations from which samples are drawn. If you considered these results in deciding whether or not to use the competency test for screening applicants in your school system, however, you would be assuming that it is reasonable to make inferences to the population of first-year teachers you might hire.

2. The null hypothesis in this example is that the average teaching effectiveness rating by principals is exactly the same, for the population of first-year teachers who score between zero and 50 on the competency test and the population of first-year teachers who score between 51 and 100 on the competency test. Another way of stating the same thing is to say that the difference between average principals' teaching effectiveness ratings is zero, for the population of teachers who have competency test scores between zero and 50 and the population of teachers who have competency test scores between 51 and 100.

3. The alternative hypothesis in this example is the converse of the null hypothesis: that the average principals' effectiveness rating of teachers who score between zero and 50 on the competency test differs from their average effectiveness rating of teachers who score between 51 and 100.

4. The risk of making a Type I error in this experiment is one percent, or .01. We know this because the footnote to the data table reads "Significant at the .01 level." Although statistics books always advise that the researcher establish a Type I error level (α) prior to conducting an experiment, that isn't always done in practice. As often as not, the researcher will conduct an experiment, compute a sample statistic, and compare it with critical values that correspond to various alpha levels. The smallest alpha level at which significance was reached will then be reported.

5. The statement that follows the asterisk at the bottom of the

table implies that the null hypothesis of no difference between population averages was tested using a Type I error level of .01 (or one percent), and that the null hypothesis was rejected in favor of the alternative hypothesis.

6. The data presented in the table support the researcher's contention that the competency test measures teaching effectiveness, but they certainly don't provide absolute proof. It is clear that those who earned competency test scores in the upper half of the range (between 51 and 100) received teaching effectiveness ratings that averaged 17.8 points higher than those received by teachers who scored in the lower half of the competency test range (between zero and 50). The finding of significance at the .01 level assures us that a difference in sample average ratings as large as 17.8 points would have occurred less than one time out of 100 (.01 = 1/100), if the averages for the populations had been equal. Thus we can be reasonably certain that principals give higher ratings to teachers who score in the upper half than they do to teachers who score in the lower half of the range on the competency test. However, we have to ask whether the principals' ratings are valid. That is, are teachers who receive higher ratings from principals really more effective teachers? In what ways are these teachers more effective? If you give it a little thought, you will see that the question of test validity is very complicated, and that it's not answered by the results of a single experiment.

7. If you recommended the use of the competency test in your school system on the basis of the validity analysis reported in the example, you would be making several inferences. First, you would be assuming that the relationship found for first-year teachers in the South Carolina school system would also be found in your school system. Second, you would be assuming that the principals in the South Carolina school system were able to identify which teachers were truly more effective. You would be accepting the validity of the principals' ratings, and the validity of the competency test.

Example B.

1. Each of the four null hypotheses tested in this example concerns the correlation between teachers' annual salaries and

the average achievement of their students. The first null hypothesis tested is that, for the population of elementary school teachers in the California school system, the correlation between teacher's annual salary and students' average reading achievement equals zero. That is the same as hypothesizing that teachers' salaries and students' reading achievement are not linearly related. The other null hypotheses also state that correlation coefficients equal zero. Respectively, they assert that teachers' salaries are not correlated with students' spelling achievement, arithmetic concepts achievement, and arithmetic computation achievement. Again, correlations of zero are hypothesized not for the sample of 100 teachers, but for the entire population of elementary school teachers in the large California school system where the research was conducted.

2. The alternative hypotheses are all of the same form. They state that, for the population of elementary school teachers in the large California school system where the research was conducted, the correlations between teachers' salaries and students' achievement (in reading, spelling, arithmetic concepts, and arithmetic computation respectively) are not equal to zero.

3. The researcher is attempting to generalize to the population of elementary school teachers in the school system from which the sample was selected. In discussing her results, however, it would not be unusual (although not absolutely correct) for the researcher to sound as though she is attempting to generalize to all elementary school teachers in the nation. The reader has to be cautious in accepting a researcher's or evaluator's conclusions. Overgeneralization in research and evaluation reports is unfortunately common. As an informed reader, you must pay attention to the population that was actually sampled, and be cautious in generalizing beyond that population to the particular population that might be of interest to you.

4. The population parameter in the hypothesis test concerning the relationship between teachers' salaries and students' reading achievement was the population correlation coefficient between these two variables; it was the correlation for the population of all elementary school teachers in the California school system and the students they taught.

5. The statement "significant, alpha = .05," the footnote with two asterisks at the bottom of the data table, refers to the correla-

tion between teachers' salaries and students' arithmetic computation scores. It means that the null hypothesis that teachers' salaries and students' arithmetic computation scores have a correlation coefficient of zero would be rejected, if tested at a Type I error level of five percent. The hypothesis test refers to the correlation between these variables for the population of all elementary school teachers and their students in the large California school system.

6. It appears that teachers' salaries are related to students' achievement test scores in some subjects, but not in others. Because there are no asterisks beside the correlations for reading scores and arithmetic concepts scores, we can infer that these correlations were not significant at the .10 level. In other words, when the researcher tested the null hypothesis that, for the entire population of elementary school teachers and their students, these types of achievement were not correlated with teachers' salaries, the null hypotheses were retained. In contrast, the null hypotheses that teachers' salaries were uncorrelated with spelling achievement and arithmetic computation achievement were rejected. This suggests that it is unlikely that the population values of the correlations between teachers' salaries and students' achievement in spelling and arithmetic computation are equal to zero. It does not tell us how large the correlations are likely to be for the entire population, however. The correlations could be so small that they are trivial, even though they are different from zero.

7. These results certainly don't suggest that raising teachers' salaries is a good way to increase students' achievement. In the first place, just because two variables are correlated, we cannot state that one variable causes the other. That is, we don't know whether high-achieving students are assigned to older (therefore better paid) teachers, or whether highly paid teachers are more effective, and therefore produce students with higher achievement. It could also be the case that teachers' salaries and students' achievement are correlated with each other only because they are both correlated with some third variable. For example, it could be that students from wealthier homes have higher achievement scores, and are assigned to schools with higher-paid teachers. If this were the case, economic status would be responsible for the correlation, so that paying teachers more

probably wouldn't affect student achievement at all.

O.K. Those are our answers. Compare yours with ours and see how you made out. We gave you fourteen questions in all. If you answered at least ten of the fourteen correctly on your first try, we would think you were doing very well, and that you had a pretty firm grasp of the logic of hypothesis testing. In case you haven't noticed, the material in this chapter isn't your ordinary bedtime reading. It's heavy and complicated, so you can be very pleased with yourself. If you didn't quite make our somewhat arbitrary cut score, don't despair. Give it another shot. First, check your understanding of the vocabulary at the beginning of the chapter. Finally, take the quiz again. When you finish, be sure to compare each question with your answer and our answer. Note the similarities and differences. If you haven't realized it already, there's a good bit of new information in our answers (sneaky devils, aren't we?).

This chapter ends our introduction to the theory of statistical inference. You should have enough vocabulary and sufficient understanding of basic concepts by now to get into interpretation of real research and evaluation reports. In the chapters that follow, we will try to help you interpret and fully understand the results sections of a number of real evaluation and research reports. These examples differ in the types of statistical inferences they report and the types of parameters involved in the inferences. They constitute a carefully selected sampling from the files of the Educational Research Information Centers (ERIC). Once you learn to interpret the results presented in the remaining chapters of the book, you'll be able to read and understand a very large percentage of the quantitative research and evaluation literature. You'll be among a small minority of administrators who can truthfully make that claim.

Problems—Chapter 8

8.1 In a report on performance on the Verbal Subtest of the Scholastic Aptitude Test (SAT), Psycho and Metric (1976) stated: "Those of high socio-economic level had a mean score of 590, while those of low socio-economic level had a mean score of only 480. The difference of 110 points was significant at the .05 level."

Based on this report, which of the following statements are true? (Circle the letter of *each* true statement.)

a. One can be 95 percent certain that socio-economic level is one of a number of causes of SAT performance.

b. In comparing populations of students of low and high socio-economic level, one can be 95 percent confident that the difference between their average SAT scores would be at least 110 points.

c. Assuming the students tested were randomly sampled from their respective populations, it is very unlikely that the average SAT score of the population of low socio-economic-level students is the same as the average SAT score of the population of high socio-economic-level students.

d. In this study, the authors apparently tested the null hypothesis that populations of low socio-economic-level and high socio-economic-level examinees had equal average SAT scores, using a Type I error probability of 0.05.

8.2 Aged and Wise (1980) reported on a study of the relationship between age of school building and the dropout rate in a sample of 76 elementary schools. They found a correlation of 0.21, which was "significant at the 0.10 level."

a. To what population might the authors be generalizing?

b. What null hypothesis are the authors testing?

c. What alternative hypothesis is implied by the authors' statement?

d. Can you be certain that a correlation of +0.21 would be significant at the 0.05 level?

8.3 The Gallup Poll recently reported that 37 percent of the movie-going public was between the ages of 14 and 18, whereas only 24 percent was between the ages of 19 and 25. As is typical of Gallup's work, this report was based on a survey of people in both age groups. In a test of the null hypothesis that equal proportions of the populations of both age groups attended movies, it was found that the test statistic fell into the critical region (Alpha equalled 0.01).

    a. What was the alternative hypothesis in this study?
    b. Should the null hypothesis be rejected or retained?
    c. Was the test statistic significant at the 0.05 level?

8.4 Think of an evaluation study involving two competing curricula, where the objective is to maximize student motivation. Suppose that you can conduct the study using random samples of students selected from a particular school system, and that a good method of measuring student motivation is available to you.

    a. State an appropriate null hypothesis.
    b. State an appropriate alternative hypothesis.
    c. Describe a Type I error in the context of this study.
    d. Describe a Type II error in the context of this study.

9.

Inferences Involving Averages

The Path Ahead

This chapter is a prototype for all that follow. In this chapter we'll examine research and evaluation studies in which the focus of inference is two population averages. In subsequent chapters, we'll examine studies in which inferences focus on standard deviations, correlation coefficients, or more than two averages. When more than two averages are involved, the appropriate statistical procedure is called the analysis of variance. Analysis of variance (ANOVA) is of sufficient importance to rate a chapter all its own. In fact, we think so highly of it, we're going to give it two. We'll then complete this marvelous tome with an investigation of studies involving a couple of the fancier statistical procedures you're likely to run into in your wanderings through evaluation and research reports: regression analysis and factor analysis.

Our goal in these chapters is not to expose you to more statistical theory (the chapters you've already plowed through on your way to this one gave you well enough of that). Rather, we hope you'll gain facility in reading and interpreting the statistical portions of evaluation and research reports by studying a series of structured examples.

The bulk of each chapter contains excerpts from real evaluation and research reports. We begin each chapter by discussing the types of inferences it treats, and then we move immediately to presentation of a number of studies that involve those types of inferences. Each study begins with a synopsis of its purposes and an overview of its design. We then present data tables and other statistical results from the author(s) original work, together with the author(s) interpretation of those data. At this point, we'll help you to see the precise inferential process the author(s) used—what the population is, what the parameters are, the hypotheses or confidence intervals involved, and the level of significance or level of confidence the author(s) used. Finally, we'll interpret the data, help you to see the ways in which the study is weak or strong, and discuss the assumptions you'd be making were you to apply the study's results to your own local setting. We know that sounds like a heavy load. But it's like olives and oysters. Try one and you'll see they're not so bad. Try two and you might even like them.

This chapter contains examples of inferences involving two population averages. In the discussions that follow you will see that two basic research questions are usually addressed in these studies. The researcher is either concerned with the difference between the averages of two populations or with the difference between the averages of a single population that exist on two different occasions (such as before treatment and after treatment). The inferential questions that apply will be made clear in the discussion of each study.

A Final Evaluation of a Career Education Program

Purpose of the Study. Kershner and Blair (1975) conducted an evaluation of the third-year operation of a career education program veloped by Research for Better Schools, Inc. The program was intended to provide secondary-school students with cognitive skills, career experiences, and personal perspectives "which aid in the selection and pursuit of adult life goals." Various community resource sites and a large urban high school were used to provide students with instruction in career exploration and specialization, career guidance, and basic skills. The bulk of the third-year program participants were eleventh-graders, but the program also served some students in the ninth, tenth, and twelfth grades.

Evaluation Design. The evaluation was extensive, if not complex. It involved five groups of students, three of which were labeled experimental groups (groups that received some program treatments), and two of which were labeled control groups (groups that did not receive program treatments). One of the experimental groups consisted of students who were chosen during the third year of program operation to replace other students who had dropped out. All students who were available for assessment at the beginning of the school year as well as at the end of the school year completed a large number of assessment instruments. These instruments included standardized measures of career maturity, self-acceptance, and achievement, in addition to specially developed measures of career attitudes and school attitudes. Demographic information was collected only once during the year, through an additional instrument called the Student Demographic Data Questionnaire.

We will consider only a small portion of the evaluation in the present example. We will focus our attention on an assessment of the attitudes of students who had either been in the program for two years, or who had been members of a control group for a period of two years. Thirty-six experimental students who were participating in the program for their second year were assessed at the beginning and end of the 1974–75 school year. Twenty-five control-group students completed the same instruments during the same time periods. The groups were composed of "returning students," and were not constructed through random assignment. Thus it is possible that the groups differed in their attitudes and on other important variables, prior to the initiation of program treatments.

Students' attitudes were measured through an Assessment of Student Attitudes Scale (ASA), that provides scores for five educational foci: education in general, the school curriculum, school resources, school counseling, and an overall attitude toward learning environments. The attitudes of second-year experimental and control group students toward all five foci were assessed at the beginning and end of the 1974–75 school year. Results are shown in the following data table.

Authors' Data Table. Table 40 in the Kershner and Blair (1975) report (p.64) provides data on student attitudes. It is reproduced below.

Authors' Interpretation. Kershner and Blair conclude from Table 40 that there were "no pretest differences between the E2 (experimental) and C2 (comparison) groups. The posttest

Assessment of Student Attitudes
Second Year Students

(Authors' Table 40)

Administration	Subtest	Group		
		Experimental (E2) n = 36	Control (C2) n = 25	"t"
Pretest	ASA Education in General	33.53	34.88	.70
	ASA School Curriculum	33.47	35.88	1.76
	ASA School Resources	33.67	33.16	.92
	ASA School Counseling	34.28	31.44	1.19
	ASA Overall Attitude Toward Learning Environments	33.67	33.64	.02
Posttest	ASA Education in General	36.81	32.44	3.76
	ASA School Curriculum	36.78	33.08	2.16
	ASA School Resources	35.81	31.88	2.30
	ASA School Counseling	37.81	29.04	4.51
	ASA Overall Attitude Toward Learning Environments	36.47	31.48	4.99

Critical value for two-tailed "t" test, df $\geq$ 30, = 1.697

Critical value for one-tailed "t" test, df $\geq$ 30, = 1.310

analyses indicated that the E2 group had a significantly more positive attitude than the C2 group on all subtests of the ASA."

"There is overwhelming support for the hypotheses that experimental students gain significantly more positive attitudes toward learning than do the control students." (p.64).

The Nature of the Inferences. From the authors' interpretation, it is clear that some hypotheses were tested to develop the results presented in Table 40. Just what were those hypotheses? What parameters were involved? How can the results be generalized? What is the level of significance? All is about to be revealed to you!

The data in Table 40 illustrate the results of testing ten different null hypotheses against ten different alternatives. Five of the hypothesis tests involve pretest data and five are based on posttests. In each case, the authors are acting as though their experimental and control groups had been randomly sampled from a population of students who might have participated in the career education program. Although they acknowledge the nonrandom selection of students, their statistical procedures largely ignore that fact. So the population to which the authors implicitly generalize their results is the population of students who might have participated in the program. In effect, the authors are asking "What if the population of potential participants had been divided in half through some random procedure, one half being assigned to the program and one half being excluded? Would the halves differ in their pre-program attitudes, as measured by the ASA? Would the halves differ in their post-program attitudes?"

The first five null hypotheses that the authors test state that the average pretest attitudes of the population of students who might have been assigned to the program are equal to the average pretest attitudes of the population of students who might have been excluded from the program. In short, the null hypotheses assert that there are no differences between the average pretest attitudes of the halves of the student population that might have been treated and untreated. For example, the first null hypothesis (the one represented by the first line of data in Table 40) states that, at time of pretest, the average ASA attitude toward "Education in General" is equal, for the population of potential program participants and the population of potential non-participants. The last five lines of the table represent the results of testing five similar null hypotheses, at the time of posttesting.

In the footnote to their table, the authors provide critical values for two types of alternative hypotheses. The value labeled "Critical value for two-tailed 't' test" applies to alternative hypotheses of the following form: "The average attitude of potential program participants differs from the average attitude of the potential population of untreated students." The value labeled "Critical value for a one-tailed 't' test" applies to alternative hypotheses of the form: "The average attitude of potential program participants is higher (more positive) than the average attitude of the potential population of untreated students." The first kind of alternative hypothesis, where it is asserted that the averages differ without saying which average is larger, is called a **non-directional alternative hypothesis.** The second kind of alternative, where one of the two averages is asserted to be larger than the other, is called a **directional alternative hypothesis.**

In this example, the authors don't state the level of significance they are using in their hypothesis tests. However, by knowing the type of hypothesis testing procedure they used, and by referring to a set of statistical tables, we were able to determine that the critical values they gave corresponded to an alpha level of 0.10. Thus they were willing to take a ten-percent risk of making a Type I error in testing each of their null hypotheses. We determined the alpha level by looking in a table of Student's t-distribution. We knew which kind of table to look in because of the "t" statistics given in the last column of Table 40, and because of the nature of the null hypotheses tested.

In this analysis, the authors have used a statistical procedure known as Student's t-test for independent samples. It is an appropriate statistical procedure for testing null hypotheses that involve the equality of two population averages. Its logic and underlying principles are identical to those we discussed in the last chapter, although it differs from those we've examined earlier in its computational details. Basically, the procedure compares the averages of two samples that were selected independently of each other, and asks whether those sample averages differ enough to believe that the populations from which they were selected also have different averages. The more the sample averages differ, the larger the "t" statistic will be. If the "t" statistic is larger than the critical value that corresponds to the desired alpha level, the null hypothesis of equal population averages is rejected. Look at the data in Table 40 to see what we mean. In testing the null hypothesis that average ASA attitudes toward Education in General were equal at time of pretest, for the populations of potential program participants and potential control students, the authors compared the average score of 33.53 for the experimental group with the average score of 34.88 for the control group. The "t" statistic of .70 for this comparison was smaller than the directional critical value of 1.310 and was also smaller than the non-directional critical value of 1.697. Thus the null hypothesis of equal population averages would not be rejected in favor of either alternative hypothesis (directional or non-directional). In contrast, look at the comparison of average ASA attitudes toward Education in General at time of posttest. The average attitude of the experimental group has increased to

36.81, while the average attitude of the control group has decreased to 32.44. The difference in average attitudes of the two samples is 4.37 points, which yields a "t" statistic of 3.76. This "t" statistic is larger than both the directional and non-directional critical values corresponding to an alpha level of 0.10. Thus we would reject the null hypothesis that the populations of potential program participants and non-participants have equal average attitudes toward Education in General, at the time of posttest. Since we have compared the "t" value of 3.76 to critical values both for non-directional and directional tests, we could either conclude that the average attitudes of these populations differ, or that the average attitude of the population of potential participants exceeds that of the population of potential non-participants.

We have not discussed the computation of the mysterious "t" statistic in this example, nor do we intend to do so. For our purposes, and we hope for yours, detailed computations are irrelevant. However, if you need those details, they can be found in a host of applied statistics texts, such as Hays (1981), Glass and Stanley (1970), or Downie and Heath (1977).

You will encounter many t-tests in your travels through evaluation and research reports. The comparison of two population averages is undoubtedly the most ubiquitous of inferential statistical procedures. In each case, you can probably recognize the test by the inclusion of a "t" statistic somewhere in the author's data table. If the reported t-value exceeds the critical value that you will probably find at the foot of the data table, or if the t-value has at least one "*" associated with it, together with an appropriate, explanatory footnote, you can conclude that the null hypothesis of equal population averages should be rejected in favor of a directional or non-directional alternative hypothesis.

Our Interpretation of Results. The real focus of the analyses reported in Table 40 is whether or not the experimental group had more positive average attitudes than did the control group, at the time of posttesting. For each of the five attitudes tested, the difference between sample averages was statistically significant at the 0.10 level, and the difference in sample averages favored the experimental group. Whether or not the differences in sample averages were large enough to be substantively important is a question that the authors did not treat. The differences in sample averages range from 3.70 points (for the ASA attitude toward School Curriculum) to 8.77 points (for the ASA attitude toward School Counseling). These differences are large enough that we can reject a hypothesis that corresponding differences in population averages are equal to zero. But they give us no assurance that the differences in population averages would be large enough to be substantively significant. That is a judgmental question that cannot be answered through inferential statistics.

The five null hypotheses that involve pretest attitude measures were conducted solely to determine whether or not the experimental group and the control group were reasonably similar at the beginning of the school year. Since the experimental group and the control group were not selected randomly, it would be unreasonable to assume that they were chosen from populations that had identical average attitudes prior to the experimental treatment. The hypothesis tests at the beginning of the school year allow the authors to conclude that null hypotheses of equal ASA attitudes would not be rejected at time of pretest, except for attitudes toward School Curriculum. (For that one attitude, the experimental group had a significantly higher average score than did the control group at the beginning of the school year). These hypothesis tests strengthen the authors' assertion that posttest differences in average attitudes can be attributed to the career education program.

Assumptions Needed to Generalize. Were you to use these experimental results as a basis for adopting the career education program in your school system, you would be making a number of assumptions. Clearly, you would feel that the attitude scales used by the authors measured important outcomes of career education. You would also have to assume that the average differences in attitudes exhibited by the experimental and control groups at the time of posttesting were of substantive importance. Finally, you would have to assume that similar average differences in attitudes would be realized by high school students in your school system. That last assumption is, undoubtedly, the most tenuous, since it presumes that the program could be installed just as effectively in your school system as it was in Philadelphia, and that your school staff would be just as effective in running the program. However, the same kinds of questions arise whenever a previously successful curriculum is adopted in a new setting.

Relationships Between Teacher Attitudes and Teacher Behavior

Purpose of the Study. This study considers the question of how teachers' attitudes toward mathematics, toward teaching as a profession, and toward students relate to their behaviors in the classroom, as judged by students and outside observers. The author (McConnell, 1978) viewed this question as one of several links in a chain that relates teachers' attitudes to students' attitudes toward mathematics and education in general. He reasoned that the only way teachers' attitudes might be communicated to students, and thereby influence the attitudes of the students, would be through teachers' overt behaviors as they conducted their classes. By investigating the ways in which teachers' attitudes relate to their classroom behaviors, McConnell hoped to establish the first link in the overall chain of relationships.

Research Design. McConnell made use of a variety of measures of teachers' attitudes and behaviors in collecting data on a sample of 47 ninth-grade algebra teachers. One set of measures, the Teacher Opinion Inventory Scales (TOI) was used in the National Longitudinal Study of Mathematical Abilities (a nationwide study of mathematics instruction in elementary and secondary schools) in the late 1960's. That earlier study was designed to secure data on the attitudes of a nationally representative sample of teachers, a fact that is particularly relevant to the portion of McConnell's study we will discuss here.

In collecting his data, McConnell was not able to randomly sample his 47 ninth-grade mathematics teachers from some well-defined population. That's not an unusual situation, and his study should not be faulted because he was forced to conduct his research within realistic, if unfortunate, bounds. In fact, McConnell is to be lauded because he specifically examined the representativeness of his sample, a consideration that is all too frequently missing in published educational research. It is this aspect of his research that we will discuss here.

McConnell examined the representativeness of his sample of 47 mathematics teachers by comparing the averages and standard deviations of their scores on the Teacher Opinion Inventory with those of the 637 ninth-grade teachers assessed in the National Longitudinal Study of Mathematical Abilities (NLSMA).

He reasoned that the NLSMA sample was designed to represent the population of ninth-grade teachers throughout the United States and, if the data for his sample were comparable to the data from the NLSMA sample, it would be safe to generalize his results to a nationwide population of teachers.

Author's Data Table. Table 5 in the McConnell (1978) paper (p.12) contains means, standard deviations, and reliability coefficients for the seven scales that compose the Teacher Opinion Inventory. Data are presented for the 47 ninth-grade mathematics teachers in McConnell's study (in the columns headed "This Study"), and for the 647 ninth-grade teachers in the National Longitudinal Study of Mathematical Abilities (in the columns headed "NLSMA"). The entire table is presented below, just as it appeared in the original paper, even though we will examine only a small portion of the data it contains.

Author's Interpretation. Among McConnell's discussion of the results shown in his Table 5 is the following:

"The second test in checking the representativeness of this sample of teachers was comparison of the means to the NLSMA means. The NLSMA results were assumed to define the population mean and standard deviation. A Z-test... was used to test whether the sample of teachers in this study could be assumed [to have been selected] from the NLSMA grade nine population."

"Two scales had means significantly different from the NLSMA ninth grade means: TOI4, 'Nonauthoritarian Orientation' and TOI7, 'Need for Approval.' The teachers in this study scored higher on 'Non-authoritarian Orientation' and lower on 'Need for Approval.' ... The majority of tests showed that this sample could have come from the population defined by the NLSMA grade nine data."

The Nature of the Inferences. In his discussion of results, McConnell does a good job of defining the purpose of the statistical tests represented by the data shown in Table 5, but he presumes that the reader can deduce just what hypotheses were tested, and how the analyses were conducted. A little clarification might be in order.

Each line of the table reports the results of a hypothesis test involving population averages. Thus seven hypothesis tests are considered in Table 5. For each of the seven TOI scales, McConnell tested the null hypothesis that the average scale score of the

Means, Standard Deviations, and Reliabilities From the Teacher Opinion Inventory on the Sample of 43 Teachers in This Study and 637 Ninth Grade Teachers From the NLSMA Y-Population Sample

		(Author's Table 5)						
Code	Scale	This Study			NLSMA[a]			Z-test
		Mean	SD	r[b]	Mean	SD	r[b]	
TOI1	Theoretical Orientation	39.25	5.93	.66	39.58	7.16	.69	−.30
TOI2	Concern for Students	34.16	5.35	.67	33.25	6.25	.69	.95
TOI3	Involvement In Teaching	31.67	5.29	.55	31.24	6.36	.67	.44
TOI4	Non-Authoritarian	23.37	4.47	.66	20.37	5.15	.63	3.82*
TOI5	Like vs. Dislike	14.23	1.02	.08	13.77	1.80	.57	1.68
TOI6	Creative vs. Rote	42.21	4.96	.47	43.64	6.62	.67	−1.42
TOI7	Need for Approval	37.27	5.02	.63	39.27	5.76	.72	−2.28*

a Wilson, Cahen, Begle, 1968
b Reliabilities (Cronbach Alpha)
* Significant at .05 level: z (.05) = 1.96

hypothetical population from which he might have sampled the 47 teachers in his study was equal to the average scale score of the population of NLSMA teachers. If these null hypotheses were retained, he would conclude that his sample of 47 teachers could be treated as though it were drawn from the NLSMA population. Thus the parameter in each hypothesis test was the average scale score of the hypothetical population from which the 47 sampled teachers could have been selected. The sample statistic that serves as an estimate of this parameter was the average scale score of the 47 teachers. From the footnote at the bottom of the table, we can see that the level of significance (alpha) McConnell used in each of his tests was 0.05.

In the first hypothesis test, the null hypothesis was that the average score of the hypothetical population on the TOI1 scale (Theoretical Orientation) was equal to 39.58, which is the mean on that scale for the 637 teachers in the NLSMA population. The alternative hypothesis was that the average score of the hypothetical population was different from the mean NLSMA score of 39.58.

In testing this null hypothesis, McConnell conducted a z-test, a statistical procedure that is virtually identical to the one described in the first example in Chapter 8. The value $-.30$ in the column headed "Z-test" is an example of a **test statistic.** The null hypothesis would be rejected only if this test statistic fell in the critical region of an appropriate sampling distribution. Just as in the example in Chapter 8, the appropriate sampling distribution is the normal distribution, and the critical regions fall beyond z-values of plus or minus 1.96. Since $-.30$ is larger than -1.96 and smaller than $+1.96$ (that is, it lies between the two critical values), it does not fall in either critical region. Thus the null hypothesis is retained. The appropriate conclusion is that we cannot reject the hypothesis that the 47 teachers in McConnell's study were sampled from a population that has an average TOI1 score that is equal to that of the NLSMA teachers.

Notice that the Z-test value for the Non-Authoritarian scale (TOI4) equals 3.82, and that it has an "*" beside it. Since 3.82 is larger than 1.96, the difference between the average TOI4 score of the teachers in McConnell's sample and the average TOI4 score of the teachers in the NLSMA sample is significant at the 0.05 level. Thus the null hypothesis must be rejected, and McConnell appropriately concluded that his sample of teachers differed from the NLSMA teachers on this attitude scale.

Our Interpretation of Results. We feel little need to elaborate on our interpretation of the results presented in Table 5, since we agree fully with the interpretation given by McConnell. On five of the seven TOI scales, the statistical evidence supports retention of the null hypotheses that the sample of 47 teachers was drawn from a population that has average scores that were equal to those of the NLSMA population. On two of the scales, TOI4 and

TOI7, the averages of the 47 teachers differ significantly (at the 0.05 level) from those of the NLSMA teachers. In one case (TOI4) the 47 teachers have a significantly higher average (you can tell this by comparing the means given in the table, and because the number in the "Z-test" column exceeds plus 1.96), and in the other (TOI7), the 47 teachers have a significantly lower average.

By way of apology, we should note that we have not concentrated on the most interesting aspects of McConnell's study in this example. In fact, the example has little to do with the central purpose of his study. However, we'll remedy that oversight in the chapter on inferences involving correlation coefficients, when we consider the relationships among teachers' attitudes on the various TOI scales.

We have also ignored a good bit of the data shown in the author's Table 5. The numbers in the columns headed "SD" (for standard deviation) and "r" (for reliability) are useful for several purposes. We have ignored them solely because they aren't essential to this example.

Since the question of generalizing these results to your own school system is not appropriate here, we will omit such discussion.

Evaluation of a Language Education Program

Purpose of the Study. An evaluation of a compensatory language arts program called Project LEAP was completed by Kovner, Burg and Kaufman in 1976. The program served elementary-level students in the Medford, Massachusetts School System, with support provided through Title I of the Elementary and Secondary Education Act of 1965.

In the words of the program director, the basic goal of Project LEAP was "to strengthen the Communication Skills, and the emotional and academic foundations on which such skills are based, of a specified number of children falling considerably below established norms in such skills, who currently reside in the identified target school areas."

Evaluation Design. Project LEAP was evaluated by administering a number of standardized achievement tests and several measures of psychomotor development, attitude scales, and behavior assessment devices to project participants in thirteen elementary schools. Tests were administered at the beginning and at the end of the school year, in order to compare students' pretest performance with their posttest performance. This evaluation design is very weak (a topic we will discuss in a later section), but was probably the most common design for evaluating programs supported under Title I, ESEA in 1976.

We will consider a portion of the evaluation that is representative of the entire study. The authors compared the pretest and posttest performances of fifth-graders and sixth-graders on four subtests of a structural analysis (language) test, and on the total test as well. They reported test results for 86 fifth-graders and 83 sixth-graders. The subtests included recognition of base words (8 items), prefixes (11 items), suffixes (8 items), and syllable division (22 items). The total structural analysis test had 49 items.

Authors' Data Table. Table 14 in the Kovner, et al. report, containing data on the structural analysis test performance of fifth-graders and sixth-graders, is reproduced below in its original form. It appeared on page 32 of their report.

Authors' Interpretation. Kovner et al. report that children in grades five and six were tested on structural analysis in September and April of the 1975–76 school year. On the basis of the data reported in their Table 14, they conclude that "fifth graders made significant gains in each area tested. Sixth graders made significant gains in recognition of base words and in syllable division. Sixth graders, as did fifth graders, attained substantial mastery of recognition of prefixes and suffixes on the posttest." (p.30).

The Nature of the Inferences. The authors used a statistical procedure known as a **paired-difference t-test** or a **t-test for dependent groups** to test null hypotheses involving each subtest and students at each of grades five and six. As is the case, by law, in all Title I, ESEA projects, the fifth and sixth-graders treated and tested in this project were not selected randomly from some larger population of students. Nevertheless, the authors treated their data as though the tested students constituted a representative sample drawn from some population of potential project participants. Each null hypothesis tested was of the same form: that there was no difference between the average pretest and posttest scores of the population of potential project participants. The alternative hypothesis for each subtest would be the direct

Comparison of Pre- and Post-Test Results
in Structural Analysis

(Authors' Table 14)

Scores	Pretest Mean and S.D.	Post-Test Mean and S.D.	t	P
	Grade 5, N = 86			
Base Words	4.5 1.9	5.1 1.8	2.526	<.05
Prefixes	9.5 2.7	10.5 1.3	3.333	<.01
Suffixes	6.4 2.2	6.9 1.2	2.031	<.05
Syllabification	11.8 3.8	14.4 3.3	5.969	<.001
Total	32.2 7.4	36.9 5.1	5.955	<.001
	Grade 6, N = 83			
Base words	4.8 1.8	5.6 1.8	3.634	<.001
Prefixes	9.7 2.4	10.3 2.1	1.724	NS
Suffixes	6.7 1.8	6.9 1.7	0.773	NS
Syllabification	14.1 3.2	15.7 2.8	5.195	<.001
Total	35.4 6.1	38.5 5.9	4.724	<.001

NS = Not significant.

opposite of the corresponding null hypothesis: that there would be some difference between the average pretest and posttest scores of the population of potential project participants.

The paired-difference t-test gets its name from the fact that it is used to compare the average scores of samples of individuals that are paired in some way (such as siblings, mothers and daughters, etc.) or to compare the average scores of a single sample of individuals, assessed at two different times (such as pretest scores and posttest scores). This is in marked contrast to the t-test for independent groups we encountered in the first example in this chapter. In that example, as in every application of the t-test for independent groups, the two samples whose averages were compared were assumed to be selected independently.

Apart from the difference in sampling procedures, the paired-difference t-test operates just like the other hypothesis testing procedures we have discussed. The sample data are used to compute the value of a test statistic labeled "t," and this computed value is compared to some critical values to determine whether or not to reject the null hypothesis. The null hypothesis is always of the form: there is no difference between population averages. The corresponding alternative hypothesis is either non-directional (in which case the alternative is merely that there is some difference between population averages), or directional (in which case the alternative is that one of the population averages is larger than the other). An appropriate directional alternative hypothesis in this example would be that the average posttest score of the hypothetical population of potential participants is larger than that population's average pretest score.

Instead of selecting a particular Type I error level (alpha), Kovner, et al. compared each computed t-value in their Table 14 with a series of critical values in the sampling distribution of Student's t-statistic. In the column headed "P," they report the alpha level of the largest critical value exceeded by each computed t-statistic. For example, for the Base Words subtest administered to fifth-graders, the t-statistic 2.526 has an associated P of "<.05." This means that the difference between the pretest and posttest averages of fifth-graders on the Base Words subtest is "significant at the 0.05 level." In other words, if one were to select an alpha level of 0.05 a priori, and then test the null hypothesis that there was no difference between the average pretest Base Words score and the average posttest Base Words score of the population of potential fifth-grade participants, that null hypothesis would be rejected. However, the t-statistic of 2.526 is not significant at the 0.01 level. Had that alpha level been selected a priori, the null hypothesis of no pretest to posttest gain would be retained. Continuing this line of interpretation, we see that the fifth-graders' pretest-posttest averages on the Prefixes subtest were significantly different at the 0.01 level; that their averages on the Suffixes subtest were significantly different at the 0.05 level; and that their averages on the Syllabification subtest and the Total Structural analysis test were significantly different at the 0.001 level. Recall that if a statistic is significant at the 0.001 level, it is also significant at the 0.01 level and the 0.05 level. However, the reverse might not be true; a statistic that is significant at the 0.05 level might not be significant at the 0.01 level.

For sixth-graders, differences between pretest and posttest averages on the Prefixes and Suffixes subtests are marked "NS," which you learn in the footnote to Table 14, indicates "Not significant." This means that the authors did not reject null hypotheses that average pretest and posttest scores of the population of potential sixth-grade project participants were equal. The authors do not indicate the level of significance they used for these two hypothesis tests, but we can assume it was 0.05 because that is the largest value listed in the "P" column of Table 14.

Our Interpretation of Results. As we have noted above, the design of this study is extremely weak although typical of its time. It is a near-perfect example of the misuse of inferential statistics. Not that the authors necessarily violated important statistical assumptions (one can't really tell whether that's a problem, although it is doubtful), it's more the case that the statistical analyses are largely irrelevant to the important evaluation questions. True, the pretest to posttest differences for fifth-graders are all "statistically significant." But we are forced to ask "so what?" Should you therefore conclude that the project was successful? We certainly hope you won't jump to that conclusion. Look at the pretest and posttest mean scores for the Base Words subtest for fifth-graders. The average pretest score was 4.5 items correct on an 8-item test. The average posttest score was 5.1 items correct, leading to a difference in averages of six-tenths of a word. Although the difference is statistically significant, would you say that it is educationally worthwhile? And even if the average posttest score had been 7.5 words correct, indicating that many students earned perfect scores on the 8-item posttest, could you conclude that the project had been successful in teaching recognition of base words? Not necessarily! There was no comparison group of fifth-graders who received all of their school's regular instruction except for Project LEAP. You have no way of knowing what caused the fifth-graders' gain of six-tenths of an item in average Base Word test performance, and you would have no way of knowing what might have caused an even greater gain. It might have been Project LEAP. It might have been due to the students being nine months older. They might have learned about base words by watching Sesame Street or the Big Blue Marble on television. Or their regular classroom teacher might have taught them about base words, totally outside of Project LEAP. We hope you now see the greatest weakness of an evaluation design that doesn't incorporate a comparison group. Without a good bit of serious and systematic sleuthing, you can't rule out any of a myriad of rival causes of whatever gains you observe. We also hope that this example helps you to realize that the use of statistical inference doesn't necessarily make an evaluation good.

Assumptions Needed to Generalize. Were you to import Project LEAP to your school system on the basis of the statistical results reported in the authors' Table 14, you would be showing more faith than logic. You'd be assuming that structural analysis was an important educational outcome; that the subtests used in this study were valid and reliable measures of the topics identified by their labels; that the differences in pretest and posttest performances shown in Table 14 were educationally important; that the differences shown were caused by Project LEAP; and that similar differences would be realized by fifth and sixth-graders in your school system, were you to leap into LEAP. Violation of any one of these assumptions would be sufficient to send you back to the drawing board. Are you ready to LEAP? Think first.

Summary

In this chapter we've presented portions of three actual research and evaluation studies. Each example involved the use of statistical hypothesis testing to compare two population averages. In discussing these examples, we hope we've helped you to identify the statistical hypotheses involved, the logic of the statistical procedures in the context of the particular studies, and appropriate interpretations of the data presented. When you read other studies that incorporate t-tests and z-tests, you will find that their structure is virtually identical to that of the examples considered here. To be sure, the details will differ. But the fundamental logic, and much of the statistical language, will be exactly the same. We suggest that you try your hand at reading a few evaluation studies involving comparisons of averages now, while the material in this chapter is fresh in your mind. Look at the authors' data tables, and form your own conclusions on the significance of the results. Only then, would we suggest that you read the authors' interpretations. You may be pleasantly surprised by your newfound caution and expertise.

Problems—Chapter 9

9.1 In summarizing a study of the relative effectiveness of two types of product sales campaigns, an author used the following data table:

Campaign	Mean Sales	t
Radio	$120,000	
Newspaper	$ 92,000	
		4.62**

**Significant at .01 level

a. What null hypothesis was tested in this study?
b. What alternative hypothesis was tested in this study?
c. Discuss the meaning of the two "**" attached to the t-statistic of 4.62.

9.2 What is the difference, if any, between a t-test for independent samples and a paired-difference t-test?

9.3 What is the difference, if any, between a non-directional alternative hypothesis and a directional alternative hypothesis?

10.

Inferences Involving Correlation Coefficients

The Nature of the Inferences

Chapter Four describes correlation coefficients as statistics that indicate the degree of relationship between measures of two characteristics of some group of persons or objects. For example, by calculating a correlation coefficient you could learn whether students who earned the highest scores on a reading test were also likely to earn the highest scores on a mathematics test. Or whether, among American women, shoe size was consistently related to height (do women with the smallest feet tend to be the shortest, and vice-versa?). If you want to be persnickety, you'll recall that correlation coefficients don't do a good job of indicating all types of relationships among variables, they merely tell whether plotted points that represent measurements of both variables tend to fall along a straight line. Statisticians and researchers would say that correlation coefficients only reflect the degree of linear relationship between two variables.

Educational researchers and evaluators have an abiding fascination with how things go together. They're forever posing titillating questions like "How does learning relate to instruction?" or "How is intelligence related to social class?" or "How does use of repeated drill and practice relate to arithmetic achievement?" Answers to all of these questions can be found in correlation coefficients. As you might imagine, correlations abound in the educational research and evaluation literature.

In this chapter we'll concern ourselves with correlation coefficients as population parameters. In particular, we'll examine studies in which something about the value of a population correlation coefficient is inferred from the value of a corresponding sample correlation coefficient. The logic of inferences involving correlation coefficients is basically the same as that you've encountered earlier, in chapters on averages. Both point estimation and confidence intervals can be used to estimate the value of a population correlation coefficient. And hypothesis testing can be used to examine the belief that a population correlation coefficient is equal to a particular value. Formal inferential procedures also exist for examining the hypothesis that the correlations between two variables are the same, in two different populations. For example, it is possible to test the hypothesis that the correlations between scores on a reading readiness test administered in kindergarten and scores on a reading achievement test administered in the first grade are the same, for boys and for girls.

The kind of correlational inference that you'll encounter most frequently in evaluation and research reports is a test of the null hypothesis that a correlation coefficient equals zero. If the null hypothesis were true, it would mean that the variables being correlated would have absolutely no linear relationship when data were gathered from all members of the population. In virtually all research and evaluation studies where this null hypothesis is tested, the researcher fervently hopes and expects that the null hypothesis will be rejected, in which case he or she will conclude that the sample correlation used in the hypothesis test is "significant," and that there is some reasonable assurance of a relationship between the measured variables in the population to which generalizations are being made.

The following examples consider the significance of correlation coefficients in a number of research contexts. Although the statistical procedures used in each example are the same, the contexts vary enough to make interpretation of results a bit different in each case.

Are Title I, ESEA Program Characteristics Related to Student Achievement?

Purpose of the Study. Title I of the Elementary and Secondary Education Act of 1965 is the largest federally funded education program for students in elementary and secondary schools. It provides funds to support school districts' instructional programs for "educationally disadvantaged" students. Using data collected from project applications and the 1970 Census, Koffler (1977) conducted an analysis of relationships between the characteristics of Title I, ESEA projects in New Jersey, and the performances of students on state-developed achievement tests. The portion of the analysis we will consider here used school systems (LEA's) as units of analysis. That is, information on Title I projects was assembled on a district-by-district basis, and students' performances on the Reading and Mathematics Tests of the New Jersey Educational Assessment Program were averaged within school districts. Thus for each school district in the state, Koffler compiled a set of average test scores and a set of information on Title I program characteristics.

Koffler's intent was to determine whether, and to what degree, such factors as Title I cost per student, number of students per instructor in Title I projects, percent of a district's students who were enrolled in a Title I project, and average salary per Title I instructor, were related to students' performance averages on the state's basic skills achievement tests. The purpose of the analysis was to develop some tentative explanatory hypotheses for the relationships found, not to determine what "caused" students' basic skills achievement levels. The author points out repeatedly that causal inferences cannot be made solely on the basis of correlation coefficients.

Evaluation Design. This study does not represent a designed evaluation so much as the systematic analysis of data compiled in a uniform data base for all school districts in New Jersey. Because he felt that relationships between Title I project characteristics and student achievement might differ in relatively affluent districts, districts with concentrations of middle-income families, and districts with concentrations of poor families, Koffler used 1970 Census data to develop a socio-economic status (SES) index for school districts. He then classified school districts into three SES categories, and computed separate correlation coefficients for districts in each category.

We will restrict this discussion to a district-level analysis of relationships for fourth-graders. However, Koffler's report includes data for seventh-graders and an analysis of relationships using schools as the unit of analysis.

Author's Data Table. Table 6 (p.22) of Koffler's report provides correlation coefficients between Title I project characteristics, student achievement in reading and mathematics, and district-level adoption of Title I reading and mathematics programs. Each row of the table contains four correlation coefficients representing the degree of linear relationship between a Title I program characteristic (e.g., cost of Title I program per student), and the four variables: district-wide average student reading achievement, district-wide average student mathematics achievement, number of Title I reading programs adopted by the district, and number of Title I mathematics programs adopted by the district. Notice that separate correlation coefficients are reported for school districts in three SES categories, labeled "low," "medium," and "high."

Koffler's Table 6 is reproduced below, just as it appeared in his report.

Author's Interpretation. Although Koffler conducted a number of tests of the significance of the correlations he reports, his interpretations treated the reported data as though they were strictly descriptive. Giving no explicit consideration to the population represented in his hypothesis tests, Koffler concludes that certain general relationships appear to be present for school districts in all SES categories. He notes that LEA's with higher Title I staff salaries usually had students who performed less well on the state's reading and mathematics tests. His interpretation of this finding depends on an assertion that high average Title I salaries might mean that a district had employed a small Title I staff, and that the small staff was unable to meet the educational needs of the district's Title I students. Koffler also notes that student-instructor ratio is negatively correlated with student achievement, suggesting that, on average, districts with more students per instructor have lower average student achievement.

The Nature of the Inferences. Whenever you find a table of correlation coefficients like Koffler's Table 6, where some correlations have one or more asterisks beside them, it means that the author tested each correlation for significance. That is, for each correlation coefficient in the table, the author tested the null hypothesis that the corresponding correlation, in the population to which he or she wishes to generalize, is equal to zero. Unless otherwise stated, the alternative hypothesis is that the corresponding

Table 6
Fourth Grade Correlations Between ESEA Title I,
Educational Achievement and Programs Adopted/Adapted

SES	Reading Test	Math Test	Reading Programs	Math Programs
Low				
Cost per student	−0.172*	−0.203*	0.143	0.120
Salary per instructor	−0.027	−0.012	0.089	0.083
Students per instructor	−0.0006	−0.295*	0.036	0.158
Percent Enrolled Students in Program	−0.513**	−0.296*	——***	——***
Middle				
Cost per student	0.026	−0.130	−0.063	0.139
Salary per instructor	−0.260*	−0.198	−0.058	0.525*
Students per instructor	−0.168	−0.013	0.068	0.408*
Percent Enrolled Students in Program	−0.274*	0.010	——***	——***
High				
Cost per student	0.053	0.117	−0.233	0.360*
Salary per instructor	−0.362*	−0.204	−0.034	0.270
Students per instructor	−0.382*	−0.233	0.171	0.176
Percent Enrolled Students in Program	−0.288*	0.027	——***	——***

(from Koffler, 1977)

*significant $P < .05$
**significant $P < .01$
***not done; see explanation in text.

population correlation is not equal to zero.

In Koffler's Table 6 you might note that some correlations don't have any asterisks beside them, some have one, and one has two. Just because a correlation coefficient doesn't rate any asterisks doesn't mean that testing its significance was ignored. It just means that the null hypothesis that the corresponding population correlation was equal to zero was not rejected, at the level of significance indicated for a single asterisk. Thus in Table 6, the correlation -0.027 was not found to be significant at the 0.05 level; we would not reject the null hypothesis that, for low SES districts, the population value of the correlation between "Salary per instructor" and "Reading Test" [score] was equal to zero. The correlation between "Percent Enrolled Students in Program" and "Reading Test" [score] was reported as -0.513 for low SES districts. This correlation coefficient has two asterisks beside it. The footnote to the table indicates that it is therefore "significant $p < .01$." Thus if for some hypothetical population of low SES school districts, you were to test the null hypothesis that the correlation between "Percent Enrolled Students in Program" and "Reading Test" was equal to zero, you would reject that null hypothesis at an alpha level of 0.01. The results of significance tests for other correlation coefficients in the table should be interpreted similarly. No asterisks means that the correlation coefficient is not significant at the 0.05 level; one asterisk means that the null hypothesis (that the corresponding population correlation equals zero) would be rejected at an alpha level of 0.05; two asterisks means that the null hypothesis would be rejected at an alpha level of 0.01. You will find similar shorthand in many tables of correlations in a great many evaluation and research reports. The definitions of one asterisk, two asterisks, etc., might be different, however.

Our Interpretation of Results. In our view, hypothesis testing really has no place in this study. Since data were collected from all school districts in the State of New Jersey, the idea of generalizing to some larger population of school districts makes little sense to us. Although evaluation and research studies often make use of "convenience samples," that is samples of students, schools or school districts that happen to be available to the researchers, using these samples to test hypothesis makes sense only if a sample might be representative of some hypothetical population.

In this study, the author has data for the entire New Jersey population, and it is very unlikely that the population of school districts in New Jersey is representative of any larger population. Virtually all of the correlation coefficients reported in Table 6 are modest in size. Given this caveat, Koffler's interpretation of results is quite reasonable. Except in low SES school districts, there is a somewhat negative relationship between a district's average Title I salary per instructor and the average student performance on the state's reading and mathematics tests. It also appears that higher student/instructor ratios are associated with lower average student achievement, although the relationships are not totally consistent across reading and mathematics performance and the SES levels of school districts. For districts in all SES categories, it appears that lower average reading achievement is associated with a larger percentage of a district's student body being enrolled in Title I programs.

We would emphasize, as does Koffler, that these results cannot be given any causal interpretation. For example, it would be extremely risky, and quite possibly wrong, to conclude that high student/instructor ratios in Title I programs cause low student achievement, and that the way to increase student achievement is to decrease the number of students served by Title I.

Generalization of the results of this example to your locale isn't a meaningful question, so we won't treat it in a separate section. It is unlikely that the correlation coefficients found for New Jersey's school districts would be the same as corresponding correlations for districts in your state. In fact, such factors as cost per student, average salary per Title I instructor, and number of students per instructor might have very different relationships with students' achievement scores in your state. Since Koffler's study was not intended to be generalized beyond New Jersey, possible inconsistencies across states in no way diminish its value.

Relationships Between Teacher Attitudes and Teacher Behavior

Purpose of the Study. In the chapter on Inferences Involving Averages we discussed a study by McConnell (1978) that investigated relationships between teachers' attitudes and their classroom behaviors. Students and outside observers rated certain classroom behaviors of ninth-grade algebra teachers. McConnell examined the relationships between these ratings and teachers' self-reported attitudes toward teaching as a profession, toward mathematics as a subject, and toward students. His objective was to explore one portion of a theory that related teachers' attitudes to students' attitudes.

Research Design. As we stated in Chapter 9, McConnell collected data on a purposefully selected sample of 47 ninth-grade algebra teachers. He asked teachers to complete several previously-used teacher opinion scales and other attitude measures. He had two adult raters observe teachers' classroom behaviors during four 30-minute visits conducted at intervals throughout the school year. In addition, he had students rate teachers' behaviors in order to determine how they viewed their teachers' actions in the classroom.

McConnell's paper contains a large number of data tables that bear on several interrelated research questions. We will restrict our attention to two specific questions in our review of inferences involving correlation coefficients. First, McConnell was concerned with the consistency of raters' evaluations of teachers on their third and fourth visits to their classrooms. Correlations between ratings of teachers' behaviors during successive classroom observations are measures of reliability. If teachers' behaviors were inconsistent across time, or if raters were unable to evaluate teachers' behaviors consistently, it would be unlikely that their rated behaviors would correlate with their attitudes or with anything else. A measure that is unreliable is essentially useless, since it is not a consistent indicator of behavior nor will it relate consistently to any other measure. In short, a measure that does not correlate with itself cannot correlate with anything else. McConnell was therefore wise in investigating the reliability of his teacher observation scales.

The second issue we will examine lies at the heart of McConnell's research. He summarized his findings on the relationships between teachers' behaviors and attitudes in a two-page table containing only those correlation coefficients that were significantly different from zero when tested at an alpha level of 0.05. You will see that he found a large number of statistically significant correlations.

Author's Data Tables. Table 2 (p. 4) in the McConnell paper contains correlations between the averages of two rater's scores for each teacher on their third and fourth visits to the teachers' classrooms. Although 47 teachers participated in some phases of his study, McConnell obtained behavior ratings only for 43 ninth-grade teachers.

Correlations between selected teacher behavior measures and selected teacher attitude measures appear in Table 8 of the McConnell paper. The table is quite long, and covers pages 18 and 19 of his report. In an earlier study (McConnell, 1977), the author investigated relationships between teacher behaviors and measures tudent learning and attitude change. A column headed "Import" in Table 8 summarizes some of the findings of that study, by reporting the number of times each teacher behavior measure was found to be significantly correlated with some measure of student learning or attitude change. Other columns in Table 8 provide the name of each teacher behavior measure (including the code for the variable in McConnell's coding scheme), the name of the category that contains the behavior measure, the "Source" of the measurement (either adult observers or pupils' ratings), the number and title of the associated teacher attitude measure, and the size and level of significance of the correlation coefficient between a teacher behavior measure and a teacher attitude measure.

Tables 2 and 8 of the McConnell paper are reproduced below, just as they appeared in the original document.

Author's Interpretation. McConnell's interpretation of the results presented in Table 2 is quite limited. He notes that "Difficulty of Lesson was notably unstable across the two visits. The other nine [behavioral ratings] were adequately correlated." (p. 4)

McConnell's interpretation of the results shown in Table 8 is extensive, and we will cite only a portion of it here. On page 24 he concludes:

> This study has shown that teacher attitudes are related to behavior patterns which are discernible by observers. The teachers who were more satisfied with teaching as a profession and who liked their subject more (in this case, mathematics) were rated as being clearer, more varied in presentations, more enthusiastic, more task-oriented, more indirect in the

Table 2

Correlations Between Item Scores on the Process
Rating Sheet on Class Visits 3 and 4 for 43 Classes

Item		r
PR1	Clarity	.382*
PR2	Variability	.375*
PR3	Enthusiasm	.636**
PR4	Business-like	.533**
PR5	Uses Student Ideas	.398**
PR6	Criticism	.565**
PR7	Structures Lesson	.347*
PR8	Higher Cognitive Questions	.595**
PR9	Probing Questions	.332*
PR10	Difficulty of Lesson	−.147

$*P < .05$ $**P < .01$

sense of using student ideas, more apt to ask higher cognitive questions, and less critical of students. The teachers who were more opposed to strict discipline and class control were viewed as being less clear, not providing as many chances for students to learn algebra, and providing less emphasis on logic and analysis.

McConnell suggests that his findings have implications beyond the field of mathematics, even though his study was limited to a small sample of ninth-grade algebra teachers in 13 high schools. He notes that the attitudes found to be important to successful teaching of algebra have analogues in other subject areas, and that the behaviors he found to be related to teacher attitudes are not unique to algebra teaching. Finally, McConnell notes the importance of including measures of teacher behavior in further research on the relationships between teacher attitudes and student attitudes.

The Nature of the Inferences. The inferences involving correlation coefficients that appear in the McConnell study are statistically identical to those conducted by Koffler (1977), although they differ in their substantive implications. For each of the ten behavior rating scales listed in Table 2, McConnell has tested the null hypothesis that, in a hypothetical population of ninth-grade algebra teachers, the observers' ratings of a teacher on the third visit to his or her classroom would be uncorrelated with corresponding ratings resulting from observations on the fourth classroom visit. If the null hypothesis were true for a particular rating scale (e.g., for the "Clarity" scale), it would mean that an observer's evaluation of a teacher's "clarity" would be totally unstable from one classroom visit to the next. The rating scale would therefore be valueless as an indicator of a teacher's instructional clarity. Presumably, although teachers are more or less clear from one day to the next, their relative positions on a clarity scale would not fluctuate so wildly that the clarity ranking of a teacher on one visit to his or her classroom would have no relationship to the clarity ranking of that teacher on the next visit. Were the null hypothesis to be retained, we would be forced to conclude that observers' ratings of teachers' clarity were totally unstable across time.

For each hypothesis test summarized in Table 2, the alternative hypothesis is that the population value of the correlation coefficient is not equal to zero. Unless otherwise indicated, in tests involving correlation coefficients you should always assume that the alternative hypothesis is non-directional.

Note that McConnell did not select a particular Type I error level (value of alpha) in conducting the hypothesis tests reported in Table 2. Rather, he reports the significance of each correlation coefficient when compared to critical values that are appropriate to two values of alpha (0.05 and 0.01). If a correlation coefficient has one asterisk beside it, you may conclude that it is significant at an alpha level of 0.05; if it has two asterisks beside it, you would reject the null hypothesis that the corresponding population correlation equals zero, at an alpha level of 0.01. The one correlation coefficient that has no asterisks (the −.147 for Difficulty of Lesson) was not found to differ significantly from zero when the null hypothesis was tested using an alpha level of 0.05. We know this because 0.05 is the largest value of alpha used in any significance test reported in Table 2.

In constructing Table 8, McConnell has selected only those correlations between rated teacher behaviors and self-reported teacher attitudes that were significant at the .05 level. In the column headed "r" he has reported the value of each sample

Table 8

Correlations Between Teacher Behavior Measures
and Teacher Attitude Measures (Significant at the .05 Level)
Grouped by Categories of Teacher Behavior for 43 Teachers

| Category | Teacher Behavior Measure | | | |
	Name	Description	Source	Import
CLARITY	PR1	Clarity	Obs.	6[a]
	RCLART	Clarity	Pupils	6
VARIETY/ VARIABILITY	PR2	Variability	Pupils	0
	RVART	Variability	Pupils	0
ENTHUSIASM	PR3	Enthusiasm	Obs.	0
	TCSZ	Exciting-Stimulating	Obs.	6
	RENTMST	Enthusiasm	Pupils	5
BUSINESS-LIKE/TASK-ORIENTED	PR4	Businesslike	Obs.	4
	TGSY	Responsible-Steady	Obs.	1
	RTASKT	Task-Orientation	Pupils	6

[a]"Import" indicates the number of significant correlations which were achieved between the teacher behavior measure and the criteria of pupil learning and attitude change in McConnell (1977).

Teacher Attitude Measure

TOI #	Title	r	Sig.
1	Theoretical Orientation	.26	.047
3	Involvement in Teaching	.44	.001
5	Like vs. Dislike	.31	.023
4	Nonauthoritarian Orientation	−.26	.047
3	Involvement in Teaching	.33	.017
5	Like vs. Dislike	.30	.035
6	Creative vs. Rote	.28	.034
—	None	—	—
2	Concern for Students	.34	.014
3	Involvement in Teaching	.48	.001
5	Like vs. Dislike	.26	.044
2	Concern for Students	.38	.024
3	Involvement in Teaching	.36	.011
5	Like vs. Dislike	.34	.013
3	Involvement in Teaching	.36	.010
3	Involvement in Teaching	.40	.001
5	Like vs. Dislike	.33	.015
6	Creative vs. Rote	.27	.030
3	Involvement in Teaching	.30	.005
6	Creative vs. Rote	.32	.018
—	None	—	—

	Teacher Behavior Measure			
Category	Name	Description	Source	Import
USE OF STUDENT IDEAS	PR5	Uses Student Ideas	Obs.	0
	RUSET	Use of Ideas	Pupils	0
CRITICISM	PR6	Criticism	Obs.	1
	RCRITT	Criticism	Pupils	0
STRUCTURES LESSON	PR7	Structures Lesson	Obs.	1
	RSTRUCT	Structuring	Pupils	0
QUESTIONS	PR8	Higher Cognitive Questions	Obs.	3
	PR9	Probing	Obs.	1
	RQUEST	Questioning	Pupils	1
DIFFICULTY	PR10	Difficulty of Lesson	Obs.	0
	RDIFFT	Difficulty	Pupils	0
OTHER	TCSC	Concrete, Practical	Obs.	0
	TCSK	Democratic, Kindly	Obs.	0
	ROPPT	Opportunity to Learn	Pupils	2

correlation for the 43 teachers observed in his study. In the column headed "Sig." he has reported the probability that a correlation at least as large as that reported in the "r" column would be found, for a sample of size 43, if the corresponding population correlation were equal to zero. Another way to interpret the numbers in the "Sig." column is as follows: They are the largest values of alpha at which we would reject the null

Teacher Attitude Measure

TOI #	Title	r	Sig.
3	Involvement in Teaching	.31	.020
5	Like vs. Dislike	.31	.022
6	Creative vs. Rote	.20	.028
—	None	—	—
1	Theoretical Orientation	−.20	.031
3	Involvement in Teaching	−.20	.025
5	Like vs. Dislike	−.41	.033
6	Creative vs. Rote	−.37	.007
2	Concern for Students	−.28	.037
3	Involvement in Teaching	.33	.033
—	None	—	—
3	Involvement in Teaching	.49	.034
5	Like vs. Dislike	.35	.001
6	Creative vs. Rote	.35	.010
—	None	—	—
—	None	—	—
—	None	—	—
3	Involvement in Teaching	.20	.033
—	None	—	—
—	None	—	—
4	Nonauthoritarian Orientation	−.20	.035

hypothesis that the corresponding population correlations were equal to zero. For example, were we to test the null hypothesis that the population correlation between observer ratings of teachers' "Clarity" and teachers' self-reported "Theoretical Orientation" was equal to zero, we would reject that null hypothesis at an alpha level of .047. Therefore, the sample correlation between these variables is significant, just beyond the .05

level (The correlation barely made Table 8, using McConnell's criterion.). In contrast, the .44 correlation between observer ratings of teachers' "Clarity" and teachers' self-reports of their "Involvement in Teaching" is significant at the .001 level. This means that the null hypothesis that the corresponding population correlation was equal to zero would be rejected, even if the researcher were only willing to run a .001 risk of making a Type I error (Note that .001 corresponds to one tenth of one percent.)

As was true for the hypothesis tests reported in Table 2, the alternative hypotheses implied by the significance tests reported in Table 8 are non-directional. Thus the alternative to each null hypothesis is that the population correlation is not equal to zero.

Our Interpretation of Results. The significance tests reported in McConnell's Table 2 are very unusual. Correlations that represent measurement reliability (called reliability coefficients) are typically treated as sample statistics, and are rarely subjected to a test of the hypothesis that their corresponding population values equal zero.

Rejection of the null hypothesis that a population reliability coefficient equals zero provides little comfort that the corresponding sample coefficient is sufficiently large. Ideally, one would like all reliability coefficients to equal one, a situation that is never realized in practice but sought nonetheless. If a reliability coefficient equals one, the corresponding measurement procedure is perfectly stable. Exactly the same result would be obtained if a person were measured repeatedly, under identical conditions, as long as that person's true status had not changed. If a reliability coefficient equals zero, measurement is totally unreliable, in the sense that the results of measurement are totally unrelated to a person's true status. If you think about these definitions, you will soon realize why you wouldn't jump for joy even if you were certain that a population reliability coefficient wasn't equal to zero

The sample reliability coefficients reported by McConnell range from barely tolerable to moderately respectable. The one negative value ($-.147$ for "Difficulty of Lesson") is totally unacceptable, of course. In this case, it is comforting that the null hypothesis of total unreliability was not rejected. The coefficients that range in the .50's and .60's are not at all unusual for short rating scales, and are quite respectable. The coefficients in the

.30's are troublesome, despite their statistical significance. Reliability coefficients limit the degree to which a scale will correlate with another variable. If measurement reliability is too low, a researcher might not be able to determine whether or not the measured variable is actually related to any other variable.

The correlation coefficients reported in Table 8, although statistically significant at the .05 level, are modest in size. They are undoubtedly limited by the reliabilities of the variables being correlated, and might represent far stronger relationships between the underlying variables. As McConnell suggests, they do support the contention that observers' ratings of teachers' classroom behaviors are related to teachers' reported attitudes. At least for the ninth-grade algebra teachers in McConnell's study, and quite likely for a larger population of algebra teachers as well, linkages between teachers' attitudes and observed behaviors are clearly established.

McConnell's suggestion that teachers' behaviors be considered in any study that attempts to explain relationships between the attitudes of teachers and students is quite reasonable, and, within the limits of generalizability of his study, well supported by his findings.

Generalizing the Results of This Study. Two factors limit the generalizability of McConnell's findings. First, his sample is composed of ninth-grade algebra teachers at 13 suburban high schools. Although McConnell compared his sample to teachers in the National Longitudinal Study of Mathematical Abilities (NLSMA), he found differences on some characteristics and did not have the opportunity to examine a number of other important characteristics. So the population to which one may safely generalize his findings remains unclear.

The second question that arises in generalizing from McConnell's results concerns the statistical treatment of his data. As is often true in correlational studies, a great many correlation coefficients were subjected to statistical hypothesis tests. If enough hypothesis tests are conducted, some significant correlation coefficients will be found, even though all corresponding population coefficients are equal to zero. For example, if 20 correlations are tested for significance at an alpha level of .05, we would expect to find that one of those correlations (five percent of the 20) was significant, even if all population correlations were equal to zero.

It is not clear from McConnell's paper how many correlation coefficients he tested for significance. In Table 8 alone, he reports 33 that are significant at the .05 level. Undoubtedly, this exceeds the number that we would expect to find just by chance, but we have no way of knowing for sure.

So in generalizing McConnell's results, you can probably conclude with a fair degree of assurance that some self-reported attitudes of ninth-grade algebra teachers are related, at least to a modest degree, to observers' ratings of their classroom behaviors. Whether these relationships generalize to teachers in other grades or to teachers of other subjects is an open question. Logic and substantive theory, rather than inferential statistics, will have to guide you in generalizing further.

Summary

In this chapter we have presented examples of one type of inference involving correlation coefficients. Although correlation coefficients are sometimes subjected to other inferential procedures, it is most common to test the null hypothesis that their corresponding population values equal zero.

We hope you will use the examples presented in this chapter as a guide to help you interpret the many similar studies you will find as you read the research and evaluation literature. Although the substance of specific inferences will differ greatly, you will find a large number of research and evaluation reports that incorporate null hypotheses, alternative hypotheses, and statistical language that are just like the ones we've discussed here. Hopefully, your journeys through the literature of your field will now be less mysterious.

Problems—Chapter 10

10.1 A study of the relationship between years of formal schooling and reading habits was conducted in the late 1970's. One of its results was a 90 percent confidence interval on the correlation between years of schooling and number of books read per year, based on a survey of 436 adults in New York state. The lower confidence limit was +0.35 and the upper confidence limit was +0.47.

a. To what population parameter does this confidence interval apply?
b. Can you conclude, on the basis of these data, that additional formal education results in adults reading more books?
c. Had the author of the study computed a 95 percent confidence interval using the same data, would it have been wider than +0.35 to +0.47, or would it have been narrower?
d. Could the author have used the same data to test the null hypothesis that years of formal schooling and number of books read per year were not correlated? Can you tell, by looking at the confidence interval, whether the null hypothesis would have been retained or rejected?

10.2 Jones and Stockton (1982) examined the relationship between teacher salaries and student reading achievement in 120 elementary school classes in North Carolina. They concluded as follows: "The correlation between teacher's annual salary and average student reading achievement in grades one through five was +0.02; this correlation was not significant at the 0.10 level."
a. What null hypothesis were the authors testing in this study?
b. Did they reject or retain their null hypothesis?
c. To what population might these results generalize?
d. What critical assumption would be necessary, to generalize these results to a larger population?

10.3 In an experiment involving quantitative aptitude, mathematics achievement, and gender, Wright and Stanley (1964) computed correlation coefficients between the mathematics aptitude scores and mathematics achievement scores of independent samples of 85 fourth-grade boys and 88 fourth-grade girls. They reported their results in the following table:

Boys	Girls
Mathematics Aptitude with Mathematics Achievement	Mathematics Aptitude with Mathematics Achievement
$r = 0.79^*$	$r = 0.62^*$

 * signif. at 0.05
**difference not signif. at 0.10

   a. To what null hypotheses does the footnote with one ''*'' refer?
   b. To what null hypothesis does the footnote with two ''**'' refer?
   c. To what populations might these results generalize?

10.4 Describe an inferential statistical study in your own context that involves at least one correlation coefficient.
   a. Define the research question you might want to investigate.
   b. Identify the population and the population parameter that are central to your research study.
   c. State a null hypothesis and an alternative hypothesis that are relevant to your study.
   d. What conclusions would you reach, were you to reject the null hypothesis?

11.

Inferences Involving Statistical Independence

The Nature of the Inferences

Suppose that a test of anxiety was administered to every member of a population of adults, and that the population was then divided into those who had low scores and those who had high scores. That is, each member of the population was placed either in a low-anxiety category or in a high-anxiety category, depending on his or her score. Next, suppose that all of the low-anxiety persons and all of the high-anxiety persons were given a test of problem-solving skill. Assume that this test was used to classify each person as a poor, average, or good problem-solver. Would the percentages of low-anxiety persons in each problem-solving category be the same as corresponding percentages of high-anxiety persons in each category? If so, we would say that, for the population of adults, anxiety and problem-solving ability were **statistically independent** variables. If not, we would say that the two variables were **statistically dependent.**

Research questions that are similar to the one posed in this example arise frequently in education and social research. The answers to such questions tell us whether knowing how a person is categorized on one variable is at all useful in predicting how they should be categorized on the other. If two variables are statistically independent, they are not related in any way at all—either linearly (as might be indicated by a correlation coefficient), or non-linearly. So that knowing how a person is categorized on one variable provides absolutely no hint as to how they might be categorized on the other. In contrast, if two variables are

211

statistically dependent, knowing a person's status on one is useful in predicting their status on the other.

The example we used to begin this chapter was decidedly unrealistic, since we assumed that we could secure measurements of the anxiety and problem-solving ability of everyone in the population. In real life, the opportunity to measure everyone in the population of interest is extremely rare. Far more often, we can obtain measurements only for some sample of persons that we hope is representative of the population of interest. Using the data gathered from our sample, we must make an inference to the population in deciding whether two variables are statistically independent or statistically dependent. These are the kinds of inferences we consider in this chapter.

When dealing with hypothesis tests involving the statistical independence of two variables, the null hypothesis is always that the two variables are independent. The alternative hypothesis is just the converse: that the two variables are statistically dependent. When the variables under consideration are categorical in nature, as is often the case, the appropriate statistical procedure is called the **chi-square test.**

When a chi-square test of independence is conducted, the statistical procedure is applied to data in the form of a **contingency table.** Contingency table is a fancy name for the simplest of two-dimensional tables. The categories of one variable form the rows of the table, and the categories of the other variable form the columns. The cells of the table merely show the numbers of sampled individuals who fall into each category of one variable and each category of the other. Not convinced that it's simple? Look at the following example. Suppose that you sampled 100 children, and it turned out that 55 were boys and 45 were girls. Let's assume that you had each child run a 100-yard dash, and that you categorized each child as fast, medium, or slow, depending on how many seconds the child took to run 100 yards. If 30 children were slow, 40 medium, and 30 fast, the resulting contingency table might be as follows:

	Slow	Medium	Fast	Total
Boys	20	30	5	55
Girls	10	10	25	45
Total	30	40	30	100

From the contingency table you can see that there were 20 slow boys and only 10 slow girls. There were 5 fast boys, and 25 fast girls. In short, the numbers in the body of the table tell you how many of the children in your sample fell into each combination of sex and speed categories.

By looking at the data in the table, it is clear that there were proportionately more fast girls than there were fast boys. For the sample, then, sex and speed for the 100-yard dash are dependent variables. Knowing a child's sex gives you information that is useful in predicting the child's running speed. The corresponding inferential question is whether the same relationship holds for the population from which the sample of 100 children was chosen. Were we to conduct a chi-square test using the data in your contingency table, we would be testing the null hypothesis that, in the population of children, sex and speed in running the 100-yard dash were statistically independent. The alternative hypothesis would be that the two variables were statistically dependent. The more dependent the two variables were in the sample, the larger the chi-square statistic would be. If the statistic was large enough, it would exceed the critical value associated with a selected alpha level, and we would reject the null hypothesis of independence.

In the examples that follow, you will see that tests of statistical independence arise in many different research and evaluation contexts. However, the data are always summarized in the same form and the resulting chi-square tests differ only in numerical detail.

Do Boys and Girls Differ in Average Intellectual Growth During Adolescence?

Purpose of the Study. From the time of Lewis Terman's famous longitudinal studies of genius in the 1930's to Hernstein's studies of the correlates of intelligence in the 1970's, it has been apparent that the IQ-test performance of adolescent boys tends to be relatively stable while the IQ-test performance of adolescent girls tends to decline. In this study, Campbell (1974) investigated the phenomenon further and explored the possibility of personality differences between girls whose IQ scores declined and girls whose early adolescent IQ scores were sustained. She was particularly interested in evidence that social expectations accounted for the IQ score decline among some girls.

Research Design. Campbell secured two sets of IQ-test scores for 471 high school seniors who attended one of two public high schools or one of two parochial schools in New York. She selected the schools purposefully, and was careful to obtain data from all sections of the state and from schools that varied in the urbanism of their attendance areas. The students for whom she obtained data volunteered to participate in her study, but she reports that only two refused to participate when asked. Although the population to which one might generalize Campbell's results is not clear from her study, the possibility of biased sampling due to the use of volunteer students would seem to be remote.

Campbell examined the school records of the high school seniors in her sample to determine their IQ-test scores during the seventh grade (her definition of the beginning of adolescence). She then had each student complete the same type of IQ test he or she had completed as a seventh-grader. Most of the students completed the Otis Quick Scoring Mental Abilities Test, and the others completed the Lorge-Thorndike Intelligence Test. In addition, Campbell had the girls in her sample complete several personality and attitude instruments.

Author's Data Table. Campbell determined the standard error of measurement of the tests used in her study by consulting appropriate test publishers' technical reports. She then classified each high school senior as an IQ gainer, an IQ loser or as a person whose IQ remained constant, depending on whether the difference between their seventh-grade and twelfth-grade IQ-test scores was, respectively, a positive change that exceeded one standard error of measurement, a negative change that exceeded one standard error of measurement, or a change in either direction that was less than one standard error of measurement.

Table 2 (p.5) of Campbell's paper reports the number and percent of young men and young women in her study who were classified as having lost, gained, or remained constant in IQ, from the seventh grade to the twelfth grade. Campbell's Table 2 is reproduced below, just as it appeared in her paper.

Author's Interpretation. In another analysis of her data, which we shall not discuss in detail, Campbell tested the null hypothesis that the average change in IQ-test score was the same, for the

Table 2

The Number of Young Men and Women Experiencing Changes in IQ Scores Greater Than the Standard Error of Measurement Over Adolescence

	Gained Gained	Lost Lost	Remained Constant Remained Constant
	No./%	No./%	No./%
Young Men	60 33.2	40 22.2	81 44.6
Young Women	56 19.3	84 29.0	150 51.7

*The differences between groups are significant:
$\chi^2 = 11.77$, df $= 2$, p $< .005$

populations of girls and boys. She used a two-sample t-test, similar to those discussed in our chapter on Inferences Involving Averages. Her interpretation of the results shown in Table 2 is combined with an interpretation of the results of the t-test.

Campbell points out that 19.3 percent of the young women gained in IQ during adolescence, while 33.2 percent of the young men gained over the same period. She also notes that 29.9 percent of the young women declined in IQ score, while only 22.2 percent of the young men experienced an IQ decline. She concludes that the patterns of IQ loss and gain for boys and girls were significantly different at an alpha level of .005. On page 5, she states:

> "It would thus appear that the first hypothesis was substantiated and that during adolescence young women declined in intellectual abilities, as measured by IQ scores, in greater numbers and to a greater degree than did young men."

The Nature of the Inference. In Table 2, Campbell not only presents sample statistics on the numbers and proportions of adolescent boys and girls who gained, lost, or remained constant in IQ-test score, she reports the results of a chi-square test of the null hypothesis that sex and change in IQ score are independent variables. The inferential results are summarized in the footnote to the table.

The chi-square statistic (represented by the Greek letter χ with a "2" one-half line above it) was reported as 11.77. Without going into the derivation of this statistic, it might be useful to outline its origins. Suppose that we had data on the sex and change in IQ of every adolescent in the population of interest. Assume further that we used these data to construct a two-way table like Table 2. If sex and change in IQ were statistically independent variables, the proportions of adolescents represented in each cell of our constructed table would have a specific relationship to the proportions represented in each row and each column of the table. In fact, the proportion in each cell could be found by multiplying the proportion in that cell's row by the proportion in that cell's column. For example, suppose that 50 percent of the adolescents in the population were boys and 50 percent were girls. In that case, the proportion in the boys' row would be .5 as would the proportion in the girls' row. If 30 percent of all adolescents had gained in IQ, the proportion in the gainers' column of the table would be .3. If sex and change in IQ were statistically independent, the proportion of boys who gained in IQ would equal .15. We found that number by multiplying .5 (the proportion of boys in the population) by .3, (the proportion of IQ gainers in the population). If the variables that define the rows and columns are statistically independent, the proportion in every cell of a two-way table constructed from data for an entire population will equal the product of the proportion of cases in the cell's row and the proportion of cases in the cell's column.

Notice that the discussion in the last paragraph assumed that we had data for every adolescent in our population of interest. In reality, we're far more likely to have data only for a sample of adolescents. We would then construct a table like Campbell's Table 2 using our sample data. Now even if the proportion of cases in each cell of a table constructed from data on an entire population was equal to the product of appropriate row and column proportions, we might not find the same relationships in a table constructed from data for a sample. You know that data fluctuate from sample to sample, and that sample results don't represent population results perfectly. The chi-square statistic compares actual sample results with what would be expected, if the variables used to construct the table were statistically independent, and the sample results represented the population data perfectly. The greater the differences between the actual sample proportions and the proportions that would be expected if the variables were statistically independent, the larger the chi-square statistic gets. The larger the chi-square statistic, the less likely it is that the variables are statistically independent. If the statistic becomes so large that the probability of having statistically independent variables is extremely small, we reject the null hypothesis of independence. Just like the statistics you've encountered in earlier chapters (e.g., the z-statistic and the t-statistic), the chi-square statistic has a known frequency distribution. It is possible to refer to this frequency distribution and define the critical value associated with any desired alpha level. If the chi-square statistic computed from the sample data is larger than the critical value associated with our selected alpha level, we reject the null hypothesis of statistical independence in favor of the alternative hypothesis that the variables are statistically dependent.

In describing the chi-square statistic and its distribution, we've oversimplified a bit. There is actually a whole family of chi-square distributions. The one we use in deciding whether to reject a null hypothesis of independence depends on the number of rows and columns in our contingency table. Each chi-square distribution is known by its **degrees of freedom.** The origin of this curious term is not important to our discussion. When a chi-square statistic is used to determine whether two variables are statistically independent, its degrees of freedom will always depend on the number of rows and columns in the contingency table. In particular, its degrees of freedom will equal (number of rows−1) times (number of columns−1). Notice that Campbell's Table 2 has two rows and three columns. Her chi-square statistic should therefore have (2−1)=1 times (3−1)=2 degrees of freedom. The part of the footnote to her table that says "df=2" indicates the two degrees of freedom.

When Campbell tested the null hypothesis that sex and change in IQ score were independent in the population from which her sample had been selected, she rejected the null hypothesis at the .005 level of significance. You can tell that by noting the "p < .005" in her footnote. To reach her conclusion, Campbell compared her computed chi-square value (11.77) with a table of critical values of the chi-square distribution for two degrees of freedom. Her value was larger than the critical value for an alpha level of .005.

As we have already noted, it is difficult to identify the population to which Campbell intends to generalize her findings. She candidly reports that schools were selected for her study, in part, because they had the sort of data she needed and were willing to participate. In addition, student participants were volunteers. Campbell selected her schools from widely varying geographic areas, different types of control (public and parochial), and different levels of urbanism. This approach ensures that generalizability will not be severely restricted, but it does not ensure representativeness. In her interpretation of results, one might infer that Campbell would like to generalize her findings either to all adolescents in New York state or to all adolescents in the United States. There is little information in her paper that clearly supports or refutes such generalization.

Our Interpretation of Results. We have already mixed editorial comments liberally into our discussions of the design of Campbell's study and the nature of the inferences she treats. There is very little to add.

Campbell's operational definitions of IQ gainers, IQ losers, and those with constant IQ scores seem quite sensible to us. It would be foolhardy to define as a "gainer" anyone whose IQ score increased at all, regardless of the size of the gain. It is well known that IQ tests are less than perfectly reliable, and that some gain or loss will occur on retesting, just due to error of measurement. By using a criterion of a gain larger than the standard error of measurement to define a "gainer," and a loss larger than the standard error of measurement to define a "loser," Campbell has appropriately handled the problem of the unreliability of IQ tests.

The use of seventh grade as an operational definition of the beginning of adolescence and the twelfth grade as an operational definition of the end of adolescence is certainly subject to challenge. The age ranges of students in these grades are fairly large. And ages of maturity vary substantially across adolescents. We cannot suggest an alternative design that would have handled these problems more effectively without being very complicated. But it would have been helpful to report the age distributions of the students when they were tested as seventh-graders and as twelfth-graders.

In a data table that we have not discussed, Campbell reports that young men gained an average of 1.62 IQ-score points and that young women lost an average of 1.33 points. This difference is fairly small, despite its statistical significance. The data in Table 2 appear more convincing to us. The relationship between sex and change in IQ score is clearly statistically significant. The sample data suggest substantively important sex differences as well. Fourteen percent more boys than girls were IQ gainers, and seven percent more girls than boys were IQ losers. These are large differences whose significance goes beyond the rejection of a null hypothesis. Causes should be sought, and if at all possible, ameliorating treatment should be applied.

We will not discuss the generalization of this study's findings to other contexts, since specific programs or treatments are not involved.

In-Level versus Out-of-Level Testing of Special Education Students

Purpose of the Study. Publishers of standardized achievement tests provide different levels of their tests for use in various grades. For example, Level 16 of the California Achievement Tests is intended for use with sixth-graders, Level 15 for use with fifth-graders, and so on. Different levels within the same test series typically incorporate subtests with the same title (e.g., Reading Comprehension, Arithmetic Computation, etc.), but the levels differ in their specific content, the difficulty of their items, and the reading ability they require.

U.S. schools commonly promote students from grade to grade on the basis of their ages as much as their abilities. As a result, many sixth-graders read no better than the average third or fourth-grader. Such students have great difficulty with levels of standardized tests intended for their grade (called in-level tests) and, in fact, do no better than would be expected had they merely chosen a response to each of the test's multiple-choice items through a random guessing procedure. In-level tests are particularly troublesome for many special education students.

Cleland and Idstein (1980) examined the use of out-of-level tests with special education students. (A test that is intended for students in a grade below the examinee's current grade is called an out-of-level test.) In particular, the authors investigated the largely untested claim that measurement validity could be increased by administering tests that were one to two grades below special education students' current grade levels. They reasoned that measurement validity would be increased if the use of out-of-level tests resulted in a larger proportion of students whose scores were above the "chance level." The chance level is the score students would be expected to earn had they guessed randomly in choosing their response to each test item.

Research Design. Cleland and Idstein administered the Reading Vocabulary, Reading Comprehension, and Math Computation subtests of the 1977 edition of the California Achievement Tests (CAT) to seventy-four sixth-grade special education students in two elementary schools. They randomly assigned students to each of six groups that differed in the levels and order of the tests administered. The first group of students received Level 16 (the sixth-grade level) of the tests, followed by Level 15 (the fifth-grade level). The second group received the same tests, but in reverse order. The third group received Level 16 of the tests, followed by Level 14 (the fourth-grade level). The fourth group received the same tests as the third group, but in reverse order. The fifth and sixth groups of students received Levels 14 and 15 of the tests. Group Five received Level 15 first, and Group Six received Level 14 first. Sixth-grade teachers administered the tests as a part of their school district's regular testing program, after receiving instructions on the nature of the experiment and proper procedures for test administration. All students who did not complete two tests in sequence were eliminated from the study.

Authors' Data Table. The authors determined that order of test administration had an effect on students' performance only for the Reading Comprehension subtest. They therefore restricted their analyses of Reading Comprehension results to students' scores on the first test of each two-test sequence. For the other two subtests (Reading Vocabulary and Math Computation) they combined data for a given test level, regardless of whether students had received that level first or second in the testing sequence.

Table 5 in the Cleland and Idstein paper (p.8) contains a contingency table for each of the three subtests used in their study. Students are classified as scoring at or below the chance level (labeled "Chance") or "Above Chance" on each of three test levels. Level "I" denotes the in-level test battery (Level 16); Level "O_1" denotes the out-of-level test intended for fifth-graders (Level 15); and Level "O_2" denotes the out-of-level test intended for fourth-graders (Level 14). Table 5 is reproduced below, exactly as it appeared in the Cleland and Idstein paper.

Table 5

Contingency Tables for Reading Comprehension, Reading Vocabulary and Math Computation at the Chance Level

		TEST LEVEL		
	I	O_1	O_2	
I Reading Comprehension				
Chance	6	5	5	16
Above Chance	17	19	21	57
	23	24	26	73
II Reading Vocabulary				
Chance	17	12	12	41
Above Chance	31	38	38	107
	48	50	50	148
III Math Computation				
Chance	2	2	4	8
Above Chance	26	26	28	80
	28	28	32	88

Author's Interpretation. Table 5 presents only one of several analyses Cleland and Idstein conducted in their investigation of the effects of out-of-level testing on the test performance of special education students. However, we will restrict our attention to their Table 5, and their interpretation of the results contained in that table. The authors state that the chi-square analyses of data in the contingency tables shown in Table 5 "were not significant for Reading Comprehension (χ^2 = .360, df 2, p>.05), Reading Vocabulary (χ^2 = 1.55, df 2, p>.05) or Math Computation (χ^2 = .707, df 2, p>.05)... Therefore, more students did not score above chance level as a result of taking an out-of-level test." In short, the authors conclude that out-of-level testing does not significantly increase the percentage of special education students who score above the chance level on standardized tests. In their summary, they advise caution in using out-of-level tests with special education students.

The Nature of the Inference. Cleland and Idstein have tested three null hypotheses in their analysis of the data shown in Table 5. First, for the Reading Comprehension subtest of the CAT they tested the null hypothesis that the percentage of students scoring at or above the chance level was statistically independent of the level of the test used. The alternative hypothesis was that scoring level (at or below chance versus above chance) and test level were statistically dependent. Second, they tested the same null and alternative hypotheses for the Reading Vocabulary subtest of the CAT. Finally, they examined the same hypotheses for the Math Computation subtest of the CAT. A chi-square test of independence was employed in testing each hypothesis. Since each contingency table in Table 5 has two rows and three columns, the appropriate degrees of freedom for each hypothesis test is $(2 - 1)(3 - 1) = 2$, as the authors note in their "df 2" statements. The "p>.05" comments in the authors' parenthetical summaries of the chi-square tests indicate that a Type I error level (alpha) of .05 was used for each test.

It is clear just by looking at the contingency tables that, for the sample of students used in the experiment, test level has little effect on the proportion of students who score at or above chance. For each subtest, the numbers of students listed in the "Chance" row is fairly constant, regardless of the level of test used. The same is true for the numbers of students listed in the "Above Chance" row of each table.

In testing their null hypotheses, Cleland and Idstein are attempting to generalize to some population of special education students. Each null hypothesis states that scoring level and test level are statistically independent variables in some population of special education students. Since Cleland and Idstein give very little information on the students used in their study, it is difficult to generalize from their findings. They tell us that students in their study were classified as "special education students," and were enrolled in the sixth grade in one of two elementary schools in the New Castle, Delaware County School District. However, they give no information on how the two elementary schools were selected, how students were selected within the two schools, or the basis for designating students as "special education students." Presumably, there is some population of special education students to which the authors' findings generalize. That population cannot be defined clearly from the information given in the paper.

Our Interpretation of Results. Cleland and Idstein have investigated an important and interesting problem. The selection of appropriate levels of standardized tests for special education students is a difficult problem facing administrators and testing coordinators in many school systems. From the data shown in the authors' Table 5, it appears that use of out-of-level tests may not be helpful.

We have already noted the difficulty in generalizing from the results of this study, and will not belabor the point further. We have one additional quarrel with the authors' methodological procedures. One of the requirements of the chi-square test is that the numbers in the contingency table represent different individuals or units. For example, the six students listed as scoring at the chance level when given Reading Comprehension Test Level I should not be the same students as the five listed as scoring at the chance level when given Reading Comprehension Test Level 0_1. For the Reading Comprehension data shown in Cleland and Idstein's Table 5, this requirement is satisfied since they only used data for the test administered first in each testing sequence. However, the chi-square analyses of data for the Reading Vocabulary and Math Computation tests are suspect because the numbers shown in different cells of the contingency tables represent the same students. For example, a given student might have scored "Above Chance" when he or she completed Level I of the Reading Vocabulary subtest, and at "Chance" when completing Level 0_1 of that subtest. Since each student's performances are reflected twice in the contingency tables for Reading Vocabulary and Math Computation, the numbers in those tables cannot come from independent samples. One of the basic requirements of the chi-square test has thus been violated, and any interpretation of the Reading Vocabulary or Math Computation data in Table 5 is suspect.

Generalizing to Your Setting. Even if the problems we have noted did not exist, several strong assumptions would be required in order to generalize the results of the Cleland and Idstein study to your own setting. You would have to assume that the sample of students used in the study fairly represented the special education students in your school system. You would be generalizing across settings, definitions of special education, and, most likely, grade levels and test batteries. The study provides no assurance that similar results would have been found for students in some grade other than the sixth, or for test batteries other than the California Achievement Tests. "Special education student" is a broad label. It means very different things in different school systems. In some states it includes gifted students, and in others only students who are mentally retarded, physically handicapped, or emotionally disturbed. Despite these limitations, the authors' recommendation of caution in using out-of-level tests with special education students is consistent with their findings. However, this study is less than definitive, and you would be well advised to seek additional evidence before concluding that out-of-level testing of special education students is unwarranted, and will typically lead to the same results as in-level testing.

A Multivariate Evaluation of an Open Education Program

Purpose of the Study. In 1975, Sewell, Dornseif and Matteson reported on the second-year evaluation of an open education program for seventh and eighth-graders. Their evaluation was extensive and complex in that it included measures of students' achievement, students' perceptions of the nature of their learning environments, and students' self concepts. In addition, the authors had teachers rate students on their attitudes, knowledge, skills, and judged sociability.

Sewell, et al. had several objectives in evaluating their open education program. First, they wanted to determine whether, and to what degree, the open education environment was regarded by students as being different from a traditional classroom environment. Second, they wanted to determine whether traditional relationships between students' background characteristics and achievement were sustained in an open education program. Finally, they wanted to compare the open education program with a traditional instructional program on a number of student outcome measures.

Evaluation Design. The open education program (named "OSCAR" by the authors) was evaluated by using a traditional experimental design. One hundred forty seventh and eighth-grade students were randomly selected to participate in the program from all six elementary schools in an Illinois school system. A parallel sample of seventh and eighth-graders was randomly selected to participate in the school system's traditional program for students at their grade levels. Both samples were stratified by grade level and sex. Each sample was composed of 35 seventh-grade boys, 35 seventh-grade girls, 35 eighth-grade boys, and 35 eighth-grade girls.

Students in the control sample followed the regular program of instruction for their junior high school. They were taught by a variety of teachers in separate classrooms for each subject in the curriculum. Their teachers were not encouraged to coordinate their instruction across subject areas. Students in the OSCAR program were taught by four teachers and two teacher aides in a single large classroom. Their instruction in language arts, mathematics, social science and science was coordinated, and often involved small instructional groups and individualization. Team teaching was encouraged.

In one part of the evaluation, teachers of both the control group students and the OSCAR students were asked to rate each student on four variables: attitude, knowledge, skills and sociability. The four teachers in the OSCAR program were asked to provide independent ratings for each participating student on each of these variables, using five-point scales. Suitable operational definitions of the attributes were provided in an attempt to increase the likelihood that each teacher interpreted each variable in the same way. Each control group student was also rated independently by four of his/her teachers on each variable. Since we will discuss some analyses involving these ratings, we must first consider the way the resulting data were treated statistically. The procedure might seem a little complicated, but it makes good sense and shouldn't be too opaque if considered one step at a time.

Whenever a group of judges is asked to rate a set of objects or persons, judges are found to differ in the standards they set. Some judges are lenient, and tend to give everyone high ratings. Others are harsh, and give low ratings to all but the most outstanding. Since they wished to compare teachers' ratings of students in the OSCAR program and the control program without the confounding effects of variation in teachers' standards, the authors transformed their rating data prior to computing any comparative statistics. For each of the four rating variables (attitude, knowledge, skills, and sociability) they first computed the average and standard deviation of each teacher's ratings of all students. They then converted each student's rating on each variable to a standard score (z-score) by subtracting the teacher's average rating from it and dividing by the standard deviation of the teacher's ratings. When this process was completed, the authors had z-score ratings of each student on each of four variables by each of four teachers. They then averaged each student's sixteen ratings to compute an overall average z-score for each student.

The median of the distribution of average z-scores was computed separately for the OSCAR students and the control group students. In each group, students whose average z-score was above the median were classified as "successful" and those whose average z-score was below the median were classified as "unsuccessful."

Once students had been classified as successful or unsuccessful, the authors conducted separate investigations, for students in the OSCAR program and the control group, of relationships between success ratings and such variables as students' sex and grade level. Their purpose was to determine whether or not the same relationships existed for both groups of students. If the same relationships were found, the authors would question the effect of the open environment that was the cornerstone of the OSCAR program.

Authors' Data Table. An unnumbered table on page 18 of the Sewell, et al. (1975) paper contains data on the numbers of students of each sex and grade level who were classified as "successful" and "unsuccessful." Separate listings of these numbers are shown in the same table for OSCAR students and control group students. The table is reproduced below, just as it appeared in the authors' paper.

Numbers of Students Receiving Mean z Ratings As "Successful" or "Unsuccessful"

	OSCAR		Control	
	Male	Female	Male	Female
7th Grade Successful	16	14	11	21
Unsuccessful	15	21	16	11
8th Grade Successful	7	20	8	14
Unsuccessful	15	6	16	11

Authors' Interpretation. The authors state that they tested the significance of grade and sex on teachers' assignments of students to "successful" and "unsuccessful" categories through a series of chi-square analyses. They performed these analyses separately for each sex, grade level, and each group (OSCAR and control).

The authors' interpretation of the tabled data appears on page 18 of their paper:

"In OSCAR, sex alone is not a significant contributor to 'success' ($\chi^2 = 1.72$, df = 1, p > .05); nor is grade level alone ($\chi^2 = 1.28$, df = 1, p > .05). Considered simultaneously, however, sex and grade level are significant predictors of 'success' ($\chi^2 = 10.32$, df = 3, p < .02)... As the data of the preceding table show, OSCAR seventh grade girls are disproportionately 'unsuccessful' and eighth grade girls are disproportionately 'successful'; somewhat opposite trends hold for the seventh and eighth grade boys, but to a lesser extent."

"In Control, however, the simple relation of sex to 'success' is significant ($\chi^2 = 6.26$, df = 1, p < .02), while grade level is not ($\chi^2 = .92$, df = 1, p > .05). In this group, boys of both grade levels are disproportionately 'unsuccessful' while girls are disproportionately 'successful.'"

In a later section of their paper, the authors conclude as follows (page 23):

"That the two programs differ is also attested by the differing relationship of teachers' ratings to student characteristics in the two programs. In the traditional program, this relation is fairly simple; girls are significantly more likely to receive higher ratings than boys. In OSCAR however, ratings of girls appear to increase substantially between the seventh and eighth grades, while ratings of boys show complementary decreases . . ."

The Nature of the Inferences. As in the studies we have already discussed in this chapter, Sewell and his colleagues were interested in the independence of variables that formed the rows and columns of a contingency table. They used the data on teachers' ratings of students to test a number of hypotheses on the relationships between students' sex and grade level and their classification as 'successful' or 'unsuccessful.'

The authors' conclusion that "In OSCAR, sex alone is not a significant contributor to 'success' . . ." is based on their test of the null hypothesis that, in the population of students who could have been assigned to OSCAR, sex of student and students' classification as "successful" or "unsuccessful" are independent variables. To test that null hypothesis, the authors combined the data for seventh-graders and eighth-graders assigned to OSCAR, forming a contingency table with two rows (one for "successful" and the other for "unsuccessful" students) and two columns (one for males and the other for females). Since the appropriate degrees of freedom for the chi square test equals (number of rows $- 1$) $\cdot$ (number of columns $- 1$), the appropriate value in this case is $(2 - 1) \cdot (2 - 1) = 1 \cdot 1 - 1$, as Sewell, et al. report ("$df = 1$"). When they tested their null hypothesis against the alternative hypothesis that students' sex and success classification were dependent variables, the resulting chi square value was 1.72. This value was less than the critical value corresponding to an alpha level of .05, in a chi square distribution with one degree of freedom; hence the conclusion that the relationship between the two variables was "not significant."

The next null hypothesis tested was that, for the population of students who could have been assigned to OSCAR, grade level and success classification were independent variables. The authors found a chi square value for this test (1.28) that was not significant at an alpha level of .05. To conduct this test against the alternative hypothesis that grade level and success classification were dependent variables, the authors combined the data for male and female students assigned to OSCAR, and once again formed a contingency table with two rows (seventh grade and eighth grade) and two columns ("successful" and "unsuccessful"). The computed chi square value was compared with a critical value from the chi square distribution with one degree of freedom.

The last null hypothesis the authors tested using the OSCAR data stated that, for the population of students who might have been assigned to OSCAR, the variables sex of student, grade level of student, and classification of students as "successful" or "unsuccessful" were statistically independent. To test this null hypothesis against the alternative that the variables were dependent, Sewell, et al. used the entire OSCAR contingency table. The table has four rows (for grade level and success classification) and two columns (for sex). The degrees of freedom appropriate to the test was therefore $(4 - 1) \cdot (2 - 1) = 3 \cdot 1 = 3$, as the authors report ("df $= 3$"). In this test, a chi square value of 10.32 was found. This value was larger than the critical value for an alpha level of .02 in the chi square distribution with three degrees of freedom. Therefore, the null hypothesis of independence was rejected. The appropriate conclusion is that, for the population of students who might have been assigned to the OSCAR program, success classification could be predicted (to some extent) by knowing students' sex and grade level.

The three null hypotheses that were tested for the OSCAR students were also tested for the population of students who might have been assigned to the control program. For control group students, the authors rejected the null hypothesis that sex and success classification were independent variables. They used a Type I error level of .02. Using a Type I error level of .05, the authors did not reject the null hypothesis that grade level and success classification were statistically independent variables. Presumably, the null hypothesis that sex, grade level, and success classification were independent variables was not rejected for the control group students.

As is the case in a number of research reports, Sewell, et al. do not specify a particular alpha level at which they tested all of their null hypotheses. One must infer from their report of results that .05 was used as a threshold for classifying a relationship as "statistically significant" or "insignificant." When the computed value of a test statistic exceeded the critical value corresponding to an alpha level of .05, the authors reported the smallest alpha level at which that value would be statistically significant. The educational research and evaluation literature contains many studies in which the results of hypothesis tests are treated in this manner.

Our Interpretation of Findings. This study is decidedly more complicated than the others we have discussed in this chapter. The evaluation question that the authors investigated required several hypothesis tests and a critical comparison of the results of all of them. In particular, the authors were not interested primarily in whether or not classification of an OSCAR student as "successful" or "unsuccessful" could be predicted from the student's sex or grade level. Rather they wanted to compare the patterns of predictability for OSCAR students and control-group students. Had they found sex and grade level to be independent of success classification for students in both populations, or that success classification was dependent upon sex and grade in both populations, they might have concluded that the OSCAR and control group environments were similar in their effects. Since they found different patterns of relationship for students in the two populations, the authors concluded that OSCAR and the control group environments produced differing results.

You might well wonder at this point: "So what?" That's a valid question for a decision-maker to pose. Realize that we have examined only a small portion of the results that Sewell, et al. present in their paper. They also explored more-traditional questions on the achievement effects of the OSCAR program. In addition to conducting typical outcomes analyses, it is often important to gain understanding of the processes that produce the outcomes in educational projects and programs. The questions that we have discussed really seek evidence on differences in the functioning of the instructional environments present in the OSCAR program and the control program. Had no differences been found, one could not rightly conclude that there were no differences in program operation since many important process variables were omitted from the analyses. However, some disappointment and suspicion would have been warranted. Having found some differences, the authors cannot rightly conclude that OSCAR presents an environment that is radically different from that of the conventional program. They have merely discovered some patterns that bear further investigation. The real key to understanding the functioning of "open" classrooms will be to learn what features of the environment, in combination with various student attributes, predict student success.

Summary

We have examined three research and evaluation studies in this chapter that were concerned, at least in part, with the statistical independence of two or more variables. In each of these studies, the authors displayed their sample data in the form of one or more contingency tables (two-way tables in which one variable formed the rows of the table and the other formed the columns). A chi square test was used to arrive at a decision on whether the variables were statistically independent or dependent in the population from which the sampled persons had been selected.

As you read the research and evaluation literature in your area of specialty, you will find many studies that are similar to those we have discussed here. Whenever you see a contingency table together with associated chi square values, you can assume that the author has conducted a test concerned with statistical independence. Hopefully, by reading the paper or report, you will not only be able to determine what was done, but whether or not it was appropriate. We also hope you will be able to discern whether or not the author's interpretation of results is sensible and consistent with the supporting data. Once again, we wish you happy hunting.

Problems—Chapter 11

Jefferson and Jackson (1842) conducted a study of the relationship between voters' political party affiliation and their position on extension of the Voting Rights Act of 1834. They surveyed a random sample of 100 Republicans, 100 Democrats, and 100 Independents, and reported the following results:

	Republicans	Democrats	Independents
In Favor of the Voting Rights Act	30	90	80
Opposed to the Voting Rights Act	70	10	20

*Chi-square$_{(2 \text{ d.f.})}$ = 93.0; signif. @ .05 level.

11.1 What null hypothesis were Jefferson and Jackson testing?

11.2 To what population were Jefferson and Jackson attempting to generalize their results?

11.3 Assuming that they acted in accordance with their findings, did Jefferson and Jackson retain or reject their null hypothesis? Support your conclusion by citing relevant data.

11.4 In a study on the relationship between academic major and attitudes toward education, Vetters and Clark (1980) administered a standardized attitude scale to independent samples of 50 students with each of four academic majors: sociology, psychology, education, and political science. They classified each student as having a "positive attitude," "neutral attitude," or "negative attitude." In reporting their results, the authors stated: "The chi-square on 6 d.f. equalled 7.04, which was not significant at the 0.10 level."
   a. What null hypothesis were the authors testing in this study?
   b. What decision did the authors reach on retaining or rejecting their null hypothesis?
   c. Making assumptions that you consider to be appropriate, interpret the results of this study in your own words.

12.

One-Way Analysis of Variance

The Nature of the Inferences

In Chapter 9, on Inferences Involving Averages, we discussed a number of research studies and evaluations in which the investigators tested the null hypothesis that two populations had the same average. In this chapter we'll also consider studies in which the investigator wants to compare population averages. However, the statistical procedures used in these studies permit more than two population averages to be compared at the same time.

As an example, suppose that you were considering the installation of a "metric education" program in a school system. Three commercial text publishers had tried to convince you that theirs was not only the most effective program, but the flashiest and most interesting as well. When you compared costs, you found that all three programs were competitive, so that effectiveness would have to be the deciding factor in selecting one of the three. Since metric education was to be introduced in grade four, you wanted to select the best program for all fourth-graders. You defined "best" as the program that would produce the highest average score on a test of knowledge of the metric system. The average was to be based on the performance of all fourth-graders in the school system.

In this hypothetical example you are really interested in the relative sizes of three population averages. Let's label the three metric curricula "A," "B," and "C" (Isn't that novel and creative of us?). Now you really want to know what the average score on the test of metric knowledge would have been if the entire population of fourth-graders in the school system had been taught using Curriculum A; what the population average would have been had they been taught using Curriculum B; and what the population average would have been as a result of using Curriculum C. Presumably, you would select the curriculum that resulted in the highest average fourth-grade score on the test of metric knowledge.

One statistical approach to this problem is to hypothesize that the population averages would have been identical, regardless of the curriculum used. The logical alternative to this hypothesis is that the three population averages differ. The statistical procedure used to test the null hypothesis of identical population averages against the alternative hypothesis that the population averages differ is called the **Analysis of Variance,** often abbreviated ANOVA. Although ANOVA is intended for use in testing hypothesis involving three or more population averages, it is also an alternative to the t-test for independent samples, when testing the null hypothesis that two population averages are equal. You will encounter several examples of the application of analysis of variance to two populations later in this chapter.

Many evaluation reports and research papers discuss problems that are similar in form to our hypothetical example on metric education. The number of curricula to be compared (or, more generally, the number of population averages to be compared) might differ, but the general structure of the problems will be virtually identical. The null hypotheses will state or imply that some number of populations have identical averages. The alternative hypotheses will state or imply the inequality of at least two population averages. In testing the null hypothesis against the alternative, the researchers will proceed as though they had collected data on random samples of individuals, selected from each of the populations of interest. Using ANOVA, the variation in the averages of these samples, from one sample to the next, will be compared to the variation among individual observations within each of the samples. A statistic termed an **f-ratio** will be computed. It will summarize the variation among sample averages, compared to the variation among individual observations within samples. This F-statistic will be compared to tabulated critical values that correspond to selected alpha levels. If the computed value of the F-statistic is larger than the critical value, the null hypothesis of equal population averages will be rejected in favor of the alternative that the population average differ.

In the examples that follow, we will describe the null hypotheses and alternative hypotheses in detail, and give considerable attention to the methods the authors used in deciding whether or not to reject their null hypotheses of equal population averages.

In-Level Versus Out-of-Level Testing of Special Education Students

Purpose of the Study. In the preceding chapter (on Inferences Involving Statistical Independence) we discussed a study by Cleland and Idstein (1980) on the use of out-of-level testing with special education students. The authors used several types of statistical analysis, including one-way analysis of variance, in investigating their research questions. We will therefore return to their study here, to consider their exploration of the hypothesis that special education students will demonstrate the same average score, regardless of the level of test used to measure their achievement.

To summarize briefly, the purpose of the Cleland and Idstein study was to determine whether the achievement of special education students could be measured more validly by administering achievement tests that were designed for students who were in lower grade levels (out-of-level tests). They reasoned that measurement validity would be greater if fewer students scored below the so-called "chance level" on out-of-level tests than on in-level tests. Additional details on the purposes of the study were provided in the preceding chapter.

Research Design. We will review Cleland and Idstein's research design briefly here, and suggest that you refer to the chapter on Inferences Involving Statistical Independence for additional details. The authors tested 74 special education students from two elementary schools in a Delaware school system. They administered three subtests of the California Achievement Tests (CAT) to each of six randomly-assigned groups of students. All tested students were sixth-graders who had completed the Reading Comprehension, Reading Vocabulary, and Mathematics Computation subtests of the CAT. Two of the groups completed sixth-grade tests and tests designed for fifth-graders; two of the groups completed sixth-grade tests and tests designed for fourth-graders; the last two groups completed tests designed for fifth-graders and tests designed for fourth-graders. The order of administration of the tests was reversed within each pair of groups. Sixth-grade teachers administered all tests as a part of the school system's regular testing program.

Authors' Data Tables. Cleland and Idstein report the results of their hypothesis tests in Table 3 of their paper. However, we will also examine data for their samples, shown in their Table 2. Table 2 reports sample averages (means) and standard deviations (SD's) of scores on all three subtests for students who completed sixth-grade levels of each subtest (labeled "I" for in-level testing), for students who completed fifth-grade levels of the subtests (labeled O_1 for the first set of out-of-level tests), and for students who completed fourth-grade levels of the subtests (labeled O_2 for the second set of out-of-level tests). These sample data will be useful in interpreting the results of analyses of variance, shown in Table 3.

Tables 2 and 3 are reproduced below, exactly as they appeared in the Cleland and Idstein paper on pages 5 and 6, respectively.

Authors' Interpretation. In their interpretation of the results shown in Table 3, Cleland and Idstein note that the "results of the analyses of variance show significant differences for reading comprehension ($F = 3.03$, df 2/143, $p < .05$) and math computation ($F = 4.86$, df 2/85, $p < .01$). Reading vocabulary showed numerical differences but these were not statistically significant ($F = 2.44$, df 2/145, $.05 < p < .10$).... In all three subtests, students tested at their assigned grade level scored numerically higher than those tested below grade level. The general trend was that the further one tested out-of-level from grade six, the lower the NCE scores became..." (pp. 5-6)

The Nature of the Inference. Table 3 summarizes the results of three separate and identical analyses of variance—one for each subtest. In each case, Cleland and Idstein tested the null hypothesis that average achievement test scores would be the same, for the populations of special education students who could have been tested using the in-level form of the test (I), the fifth-grade level of the test (O_1), and the fourth-grade level of the test (O_2). The alternative hypothesis was that the population averages would differ, for at least one pair of populations (either those

Table 2. Means and Standard Deviations for Reading Comprehension, Reading Vocabulary and Math Computation Scores

| | TEST LEVEL | | | | | |
| | I | | O_1 | | O_2 | |
TEST	MEAN	SD	MEAN	SD	MEAN	SD
Reading Comprehension	23.38	(12.97)	23.18	(11.30)	18.26	(10.84)
Reading Vocabulary	25.27	(13.52)	19.66	(12.40)	21.62	(12.21)
Math Computation	27.04	(11.19)	21.36	(11.25)	17.34	(13.38)

Table 3. Analysis of Variance for Reading Comprehension, Reading Vocabulary and Math Computation

TEST	SOURCE	SS	dF	MS	F
Reading Comprehension	Total	20,420	145	—	—
	Between Groups	830	2	415	3.03*
	Within Groups	19,590	143	137	
Reading Vocabulary	Total	24,220	147	—	—
	Between Groups	792	2	396	2.44
	Within Groups	23,428	145	162	—
Math Computation	Total	13,756	87	—	—
	Between Groups	1,408	2	704	4.86**
	Within Groups	12,348	85	145	—

* $p < .05$
** $p < .01$

tested with I and those tested with O_1, or those tested with I and those tested with O_2, or those tested with O_1 and those tested with O_2).

In our earlier discussion of this study, we noted that Cleland and Idstein were attempting to generalize their results to some population of special education students. We also noted how difficult it would be to identify or describe an appropriate population of generalization because the authors describe neither their sampling procedures nor their samples of tested students. We must be content with the conclusion that out-of-level testing appears to result in significantly lower average performance for some population of sixth-grade special education students on some subtests of the CAT. However, the nature of that population is not clear.

The largest level of significance (alpha level) Cleland and Idstein chose to regard as appropriate for rejection of a null hypothesis was .05. We can discern this from their interpretation of the Reading Comprehension results and the Reading Vocabulary results. In the former case, the F-statistic in the body of the table (3.03) has an asterisk beside it, and the corresponding footnote indicates that the probability of observing an F-statistic at least this large, given that the null hypothesis was true, is less than .05. In the interpretive statement quoted above, the authors did not regard the F-statistic resulting from the Reading Vocabulary analysis to be statistically significant, even though the probability of its occurrence was between .05 and .10. Finally, the F-statistic resulting from the analysis of the Math Computation data (4.86) was so large that it exceeded the critical value corresponding to an alpha level of .01. Once again, the footnote "**$p < .01$" provides the necessary evidence.

Without going into a great deal of detail, we think it would be worthwhile to discuss the logic of analysis of variance and the meaning of the overabundant shorthand that appears in Table 3. The reasoning behind the analysis of variance is as follows: If two or more populations have identical averages, the averages of random samples selected from those populations ought to be fairly similar as well. The sample averages probably wouldn't be identical simply because of sampling fluctuations. You know that sample statistics vary from one sample to the next, and are rarely equal in value to their corresponding population parameters. However, large differences among the sample averages would cause us to question the hypothesis that the samples were selected from populations with identical averages.

So in testing the null hypothesis that our samples had been selected from populations that had the same averages we wouldn't want to reject that hypothesis simply because the sample averages differed a little bit. The question immediately arises, how much should the sample averages differ before we conclude that the null hypothesis of equal population averages is untenable? In analysis of variance, the answer to this question is obtained by comparing the variation among the sample averages to the variation among observations within each of the samples. Only if the variation among the sample averages is substantially larger than the variation within the samples, do we conclude that the populations must have had different averages.

The various statistics reported in Table 3 result from the application of the logic we have just described. We will discuss them one at a time. Note that three sections of Table 3 are identical in format. They differ only in the values of their numerical entries, and that is because each section refers to a different subtest. Let's consider the column headings first. The column headed **source** contains three entries "Total," "Between Groups," and "Within Groups." These are three "sources" of variation in the subtest scores of all students who completed each level of the subtests. The next two columns are headed "SS" and "dF." The "SS" is shorthand for **sum of squares** and the "dF" is shorthand for **degrees of freedom**.The sum of squares is a statistic that reflects variation; the more variable a set of scores, the larger the sum of squares will be. Unfortunately, the sum of squares also depends on sample size, so that it isn't a "pure" indicator of variation. This is where degrees of freedom come in. The numbers in the dF column depend on sample size or on how many population averages are being compared, depending on whether we are talking about degrees of freedom "within groups" or "between groups." For "within groups," it's the sample size that determines degrees of freedom; for "between groups," it's the number of averages being compared. The degrees of freedom are used to adjust the sums of squares for sample size and number of groups, thereby forming **mean squares** (those are the numbers in the "MS" column). The mean squares can be compared with each other to determine whether most of the variation in the original data is between individuals in the same group, or between the sample averages. The sums of squares can't be compared directly because they depend on sample size as well as on variation.

All this might be sounding a bit thick and pointless by now, but have faith, since we're almost out of column headings. The last one in the table is "F." The number in the "F" column is called an "F-statistic." It's the test statistic used to determine whether the population averages are significantly different. In other words, the F-statistic is compared to a critical value in an appropriate table to determine whether or not the null hypothesis that all populations have the same average should be rejected. If the computed F-statistic is larger than the critical value that corresponds to a selected alpha level, the null hypothesis is rejected. If the computed F-statistic is smaller than the critical value, the decision must be to stick with the null hypothesis.

Let's discuss the numbers in the top section of Table 3. Those in the row labeled "Total" were computed using the Reading Comprehension scores of all 146 students for whom data were available, regardless of the test level the students completed. The 20,420 in the SS column represents all of the variation in the data, regardless of test level. The number in the degrees of freedom (dF) column (145 here) will always equal one less than the total sample size. The numbers in the row headed "Between Groups" result from differences between the averages of the samples being compared. The more the sample averages differ from each other, the larger will be the number in the sum of squares (SS) column. The number of degrees of freedom (dF) in the Between Groups row will always equal one less than the number of populations being compared in the study. Since three populations are being compared in this study, dF between equals 2. The number in the mean square (MS) column, 415, is just the number in the SS column, 830, divided by the number of degrees of freedom, 2. Stick with it. There's only one more row to talk about, and then you'll know what all of the numbers mean!! All of the numbers in the "Within Groups" row of the table result from variation among the scores of students who are in the same sample. In this example, the more the Reading Comprehension scores of students who completed tests of the same level differ from each other, the bigger the sum of squares (SS) within groups. For example, had there been more variation among the scores of those who completed Level I of the tests, the 19,590 in the SS column would have been a larger number. The number in the degrees of freedom (dF) column, 143 in this case, depends on the total sample size and the number of populations being compared. It always equals the total number of persons for whom data have been collected, minus the number of populations being compared. Just as was the case in the Between Groups row, the mean Square (MS) is just the sum of squares divided by the degrees of freedom (19,590 / 143). The more the scores within a sample differ from each other, the larger the MS Within Groups row will be.

We've given you a lot of facts so far, but we haven't really appealed to your logic. Here it comes. We've said that the more the sample averages differ from each other, the larger the MS Between Groups will be. Also, the more the scores in a given sample differ from each other, the larger the MS Within Groups will be. Now you might think that a large enough MS Between Groups would suggest that the samples couldn't have been selected from populations that had the same average. After all, samples drawn from populations with the same average shouldn't have sample averages that are all that different. So far, your logic is right on target. However, there's one little hitch. Suppose that the scores within each of the groups were highly variable, just because people tended to differ a lot on the phenomenon being measured (e.g., Reading Comprehension). If that were the case, the sample averages might also differ quite a bit even though the samples had been selected from populations with the same average. To guard against this possibility, we compare the Mean Square Between Groups to the Mean Square Within Groups. Only if the MS Between Groups is much larger than the MS Within Groups do we decide that the samples couldn't have come from populations with the same average. The F-statistic, as we've noted above, is just the MS Between Groups, divided by the MS Within Groups. If it's larger than a critical value that corresponds to our chosen alpha level, we reject the null hypothesis of equal population averages. You don't have to worry about computing the critical value. It can be found in a table headed **"F-Distribution,"** in the back of almost any statistics book.

We've asked you to wade through a lot of terminology and numbers in this discussion, not because we felt that it was mandatory that you understood every number in the first section of Cleland and Idstein's Table 3. We invested this much in our writing and your reading because you'll find thousands of studies that present data in tables structured exactly like Cleland and Idstein's Table 3. Their table is an example of what is termed an **analysis of variance table.** Analysis of variance tables may differ in fine detail, but they are always identical in basic structure. If you understand one, you can understand all of them. So the investment in this example is well worth our effort, and yours too. If you aren't too sure of where the MS Within Groups came from, go back and read this section again. As you try to read one of the many research and evaluation reports that use analysis of variance, some time in the future, you'll be glad you spent the time now.

Our Interpretation of Results. Apart from questions of generalizability, we agree with Cleland and Idstein's interpretation of the results shown in Table 3. For some population of sixth-grade special education students, average performances on the Reading Comprehension and Math Computation subtests of the California Achievement Tests do depend on the level of the subtests used. We have already discussed the difficulty of attempting to define that population of students, given the sparse information the authors provided on their sampling procedures and their samples.

If you will take a moment to study the data shown in Cleland and Idstein's Table 2, we think you will see why significant differences between population averages were found for the Reading Comprehension and Math Computation subtests, but not for the Reading Vocabulary subtest. First of all, notice that the difference between the largest sample average (23.38) and the smallest sample average (18.26) is more than five points for the Reading Comprehension subtest. Notice also that the standard deviations of scores within the three samples are comparatively small. The largest standard deviation is 12.95 points, but those for the other two samples are close to 11 points. Thus the sample averages differ quite a bit, and the scores within samples are relatively homogeneous. This combination of conditions produces the fairly large F-statistic shown for Reading Comprehension in Table 3 (3.03). Second, study the data for the Reading Vocabulary subtest shown in Table 2. Notice that the largest and smallest sample averages differ from each other by more than five points. You might therefore wonder why the F-statistic for Reading Vocabulary, shown in Table 3, was not larger. The answer lies in the size of the standard deviations within the three samples. All are larger than 12, and one is 13.52. These large standard deviations result in a large Mean Square Within Groups, thus negating the large Mean Square Between Groups produced by sample averages that differ from each other. The result is an F-statistic (2.44) that is smaller than the critical value for an alpha of .05. The very large F-statistic (4.86) for the Math Computation subtest results from the combined effect of having sample averages that differ from each other by as much as 10 points (See Table 2), and relatively small standard deviations within two of the three samples. The Mean Square Within Groups is therefore moderately small, and the Mean Square Between Groups is quite large.

Generalizing to Your Setting. We have discussed generalization of Cleland and Idstein's findings at some length in the chapter on Inferences Involving Statistical Independence. We suggest that you refer to that chapter for a review of the issues.

An Evaluation of an Urban Day Care Program

Purpose of the Study. This report presented the results of an evaluation of the Get Set Day Care Program that was operated by the School District of Philadelphia in the early 1970's. Since the author of the report was not identified, we will refer to the report as "Get Set (1972)."

The purpose of the evaluation was to compare the achievement of children who had participated in the Get Set program for at least two years with the achievement of children who had no previous nursery school experience. The evaluation examined just one of several potential program outcomes. The Get Set Program was comprehensive, in that it included health services, nutrition enrichment, social services, extensive parental involvement, and psychological services. A series of inservice training workshops was provided for about 300 teachers and 300 teacher aides who had no previous training in early childhood education. However, no evaluation of the effectiveness of these program components was provided in Get Set (1972).

Evaluation Design. The evaluator hypothesized that "the Get Set Day Care children with two years of previous Get Set Day Care experience would perform better on the Stanford Early School Achievement Test than the children with no previous Get Set Day Care experience in the kindergarten program." (Get Set, 1972; p. 2)

To examine this research hypothesis, the author administered the Stanford Early Achievement Test to 1391 kindergarten pupils who had been in the Get Set Day Care program for at least two years, and 702 kindergarten children who had neither Get Set nor any other nursery school experience.

The Stanford Early Achievement Test has four parts, labeled Environment, Mathematics, Letters and Sounds, and Aural Comprehension. Performance on these four parts is aggregated to form a composite score. We will consider the author's analysis of children's composite test performances.

Author's Data Table. Table 3 (p. 8) in the Get Set (1972) report provides statistics on the composite test performance of children in the author's treatment group (those with at least two years of Get Set experience) and children in the author's "control group" (those with no previous nursery school experience). In addition, Table 3 shows the results of an analysis of an analysis of variance of the composite test performance data, and a separate analysis of the significance of differences in treatment group and control group performances on Reading and Mathematics subtests. We will not discuss the Reading and Mathematics test results here. We have included them in Table 3, reproduced below, only to provide an exact duplicate of the author's table.

Author's Interpretation. The author's interpretation of the results shown in Table 3 is limited in detail but grandiose in its inferences. The author states (p. 4) that "The study indicated that Get Set Day Care experiences promote the child's ability to learn in the school environment, particularly in the areas of reading and mathematics." The conclusion of the report is as follows: "Get Set Day Care experiences help children develop concepts in learning that are basic for a child when the child enters school. The children learn reading readiness skills and mathematics skills while attending Get Set Day Care. The children have the ability to learn and are learning." (p. 5)

The Nature of the Inference. The type of inference underlying the data shown in Get Set's Table 3 is identical to those discussed in our previous chapter on Inferences Involving Averages. The null hypothesis is that the average composite score on the Stanford Early Achievement Test is the same for the population of children who could have had at least two years of Get Set Day Care experience and the population of children who could have had no prior nursery school experience. The alternative hypothesis is

**Table 3. Stanford Early Achievement Test
Composite Score Language Arts and Mathematics**

	Treatment Group Children with Get Set Day Care Experience A	Control Group Children with no Previous Get Set Day Care Experience B
Sample Size	1391	702
Mean Score	84.090	78.179
Standard Deviation	19.289	20.936

		Analysis of Variance		
	Sum of Squares	DF	Mean Square	F Ratio
Within Groups	824481.312	2091	394.300	
Between Groups	16297.824	1	16297.824	41.334
Total	840779.136	2092		p = .001

	Get Set Children Percentile Rank	Children with No Get Set Experience	Significance
Reading	62	33	.001
Mathematics	53	12	.001

that these two population averages are different.

This null hypothesis could have been examined by using a t-test for independent groups (as illustrated in the studies discussed in our previous chapter), but the author chose to use analysis of variance instead. When only two population averages are being compared, use of the t-test or analysis of variance is quite arbitrary, since the results will be exactly the same. If one procedure results in a finding of statistical significance, the other will as well. If one procedure leads to support of the null hypothesis, the other procedure will merely confirm that decision.

We have described the populations underlying the inferences as consisting of "children who could have participated in the Get Set program" and "children who could have had no prior nursery school experience." These descriptions are admittedly loose. But the design of the study does not permit greater specificity. The author has presumably administered the Stanford

Early Achievement Test to all kindergarten children in the Philadelphia School System who fell into either participation category during the 1971-72 school year. So you might think that data have been collected from all members of the populations of interest in that school year, and that there is no inference to be made. However, decisions on the efficacy of education programs should depend on their performance over a period of years, not just their apparent effects in a single year. So we might reason that the 1971-72 populations constitute random samples from populations of children who might be eligible for participation in the Get Set program in future years. The inference then applies to these larger populations.

The "p=.001" in the "Analysis of Variance" section of Table 3 indicates that the author used a very small alpha level (.001 equals one tenth of one percent) in testing the null hypothesis that the average composite scores on the Stanford Early Achievement Test were equal for the two populations. The very large F-statistic (41.334) exceeded the critical value corresponding to this small alpha level. As we noted in the previous example in this chapter, the degrees of freedom figures in this analysis of variance table can be verified by considering the number of population averages being compared and the sizes of the samples being compared. The degrees of freedom between groups is, correctly, equal to one, since there are only two populations under consideration. The total DF is, appropriately, one less than the total sample size (1391 plus 702). And finally, the DF Within Groups is just equal to the difference between the DF Total and the DF Between Groups. Note that the form of the analysis of variance table is virtually identical to that shown in the previous example.

Our Interpretation of Results. We have no quarrel with the statistical procedures used in the Get Set (1972) evaluation report. The analysis of variance appears to have been carried out correctly. However, we strongly question the author's interpretation of the study's results.

In his or her interpretation, the author not only suggests that children who participated in the Get Set program achieved a significantly higher average score on the Stanford Early Achievement Test, but that the Get Set program caused that higher achievement. That could be the case. But the evidence provided in the Get Set (1972) evaluation report is less than compelling.

You should immediately realize that the author's treatment group and experimental group were not randomly assigned. This suggests the possibility that Get Set participants were more able than their non-participant counterparts, at the time they enrolled in the program. Since Get Set is a pre-school program, it is clearly voluntary rather than compulsory. Only children with parents who care enough about their children's development to enroll them in the program and then give of their own time, constitute the treatment group. These children might well be regarded as advantaged at the outset. They might have shown superior performance on the Stanford Early Achievement Test in kindergarten just due to the support of their parents. We thus have at least one viable rival interpretation of the Get Set results. There are many others, including differences in the aptitudes of children who enroll in the program and those who do not; differences in the transiency rates of parents who enroll their children in preschool programs (i.e., it's really family stability that produces increased preschool achievement); the development of test-taking skills, rather than superior achievement, by program participants; etc.

The design of the Get Set evaluation unfortunately permits each of these rival interpretations to be entertained as viable alternatives to the conclusion that Get Set caused the apparent increases in pre-school achievement. Without random assignment of children to the treatment group and a true control group, or the use of a much stronger non-experimental evaluation design, there is no sound basis for claiming that the program resulted in greater achievement than would have been realized in the absence of the program.

Generalizing to Your Setting. Suppose that the Get Set evaluation had been designed differently, and that you could clearly attribute the superior performance of the treatment group to the effects of the Get Set program. What assumptions would be required to generalize these findings to your own setting? First, you would have to assume that the sample of children used in the Get Set evaluation were representative of the population of pre-school children in your school system. Second, you would have to assume that the Get Set program could be replicated in your school system. This would require investment of the same time and dollar resources (at least proportionately) as in Philadelphia, and the same level of commitment and capability on the part of teachers, counselors, psychologists, nutritionists and administrators as was realized in Philadelphia. Finally, you would have to be convinced that the Stanford Early Achievement Test was a valid measure of the kinds of outcomes that you would expect from a successful day care program.

Given the limitations in the **internal validity** of this study (i.e., one cannot causally attribute the results of the study to the treatment), the question of generalizability of results is moot for practical purposes. It will always be the case that internal validity is a necessary ingredient in the generalizability of research findings. If a research or evaluation study is not internally valid, generalizability of its findings is out of the question. However, even the findings of an internally valid research study might not be generalizable to your setting. You have to address both questions separately, and achieve satisfactory answers to both before you apply research findings on your own home ground.

An Experiment on Teaching Prospective Teachers to Explain

Purpose of the Study. Pool and Capie (1977) conducted an experiment to determine the effectiveness of a program designed to teach prospective teachers how to explain concepts to elementary school students. The authors created five instructional modules on explaining, based on the "How to Explain Program" developed by Miltz (1971). Each self-instructional module required 90 minutes of study and practice time. The five modules were titled (1) Listening, (2) Structuring, (3) Application / Validity / Simplicity / Clarity, (4) Focus / Rule-Example-Rule / Vagueness / Summarizing, and (5) Check on Yourself.

The purpose of the research was to determine (1) whether use of the instructional materials would result in preservice teachers achieving a higher score on a test on principles of explaining than would preservice teachers who received no instruction in explaining, (2) whether elementary pupils' average score on a concept attainment test would be affected by their teacher's having participated in the "How to Explain Program." and (3) whether there is a relationship between preservice teachers' scores on a test on principles of explaining and elementary pupils' scores on a test of concept attainment. We will consider only that portion of the Pool and Capie study concerned with their second research objective.

Research Design. The principal subjects in the Pool and Capie (1977) study were 35 prospective elementary school teachers who were in their junior year of teacher training. The researchers randomly assigned 17 of the prospective teachers to a treatment group and the other 18 to a control group. Those in the treatment group completed the five self-instructional modules on explaining that Pool and Capie adapted from Miltz's "How to Explain Program."

To determine the effectiveness of their instruction in explaining, Pool and Capie randomly assigned all prospective teachers, whether in the treatment group or the control group, to one of two elementary schools. Pupils enrolled in those schools were randomly sampled and assigned to "mini-lesson groups" containing six to nine pupils. The preservice teachers were randomly assigned to the groups and were required to teach a fifteen to twenty-minute lesson to their randomly assigned pupils. The lesson was intended to help the pupils acquire a specially constructed mathematical concept, called an "**IRT**."

A concept is a set of entities that have certain prescribed properties in common. The IRT was intentionally designed to be unfamiliar to the prospective teachers and to the elementary school pupils. All IRTs have three interconnected triangles, a circle that is divided into two equal portions, and a rectangular shape in their figures. Certain requirements on shading of the figures were also specified by the researchers. Any figure that did not meet all of the requirements for shape and shading was not an IRT. The object of the lessons was to teach the elementary school pupils to distinguish IRTs from non-IRTs. The authors provided the preservice teachers with four examples of IRTs and non-IRTs, for use in the lessons.

Pool and Capie developed a 20-item test of pupils' attainment of the concept of an IRT. The test consisted of 20 figures which were either IRTs or non-IRTs. The pupils were asked to mark "yes" if they thought a figure was an IRT and "no" if they thought a figure was not an IRT. A pupil's score on the test was the number of correctly identified figures. The test was administered to pupils in all mini-lesson groups, following the prospective teachers' lessons on IRTs and non-IRTs.

Author's Data Tables. Pool and Capie administered their concept attainment test to 262 elementary school pupils. One hundred twenty-six of these pupils were randomly assigned to groups taught by prospective teachers who had received the instruction on how to explain (Treatment Group Teachers) and 136 pupils were randomly assigned to prospective teachers who had not received the instruction (Control Group Teachers).

The authors completed a one-way analysis of variance using the concept attainment scores of both sets of pupils. Sample statistics were displayed in their Table I, and the analysis of variance results were shown in Table II. These tables are reproduced below, just as they appeared in Pool and Capie's paper (p. 9).

Table I. Comparison of Concept Attainment Means

Treatment Group	Control Group	Total
$\bar{X} = 17.418$	$\bar{X} = 16.81$	$\bar{X} = 17.104$
n = 126	n = 136	n = 262
N = 17	N = 18	N = 35

n = pupils
N = preservice teachers

Table II. Analysis of Variance on the Student Concept Attainment Measure

Source	Degrees of Freedom	Sum of Squares	Mean Square	F. Ratio
Between Groups	1	31.88	31.88	3.347*
Within Groups	260	2475.81	9.52	
Total	261	2507.69		

*probability (<.065)

Authors' Interpretation. Pool and Capie were appropriately cautious and refreshingly thorough in interpreting their results. They first noted that "Elementary pupils taught by preservice teachers receiving instruction in the 'How to Explain Program' performed better (p < .065) than pupils taught by teachers not receiving instruction" (p. 11). They then encouraged the reader to use caution in generalizing from their findings for a variety of reasons. First, the concept attainment performance of pupils taught by control group teachers and treatment group teachers was quite high. On average, both groups answered more than 75 percent of the test items correctly. The pupils used in the experiment were enrolled in grades one through five. The authors suggested that their test was not effective in discriminating between the concept attainment of fourth-graders and fifth-graders who had been taught by teachers in the treatment group and the control group. In short, the test was too easy.

The authors next noted that preservice teachers in the control group had been assigned to their instructional groups for an extra week, prior to teaching the lesson on the concept of an IRT. For a variety of rasons, including the opportunity for the preservice teachers to gain in-class experience, and the establishment of rapport between the control group teachers and the elementary pupils, the performance of pupils taught by the control group teachers might have been inflated by this design.

Finally, Pool and Capie noted that they had provided both control group and treatment group teachers with structured lesson plans. These materials probably added to the teaching effectiveness of both groups, and further diminished any differences between the groups' effectiveness. It is interesting to note that Pool and Capie found a difference between the average concept attainment scores of pupils taught by the treatment and control group teachers despite these possible design flaws. Presumably, the difference might have been greater had the control group teachers not been provided with these advantages.

The Nature of the Inference. Pool and Capie tested the null hypothesis that the population of elementary school pupils who could have been taught by teachers trained with the "How to Explain Program" materials would have the same average score on a concept attainment test as would the population of pupils who could have been taught by teachers who had not experienced such training. Since their experiment involved samples of prospective teachers and elementary school pupils, the authors intended to generalize to both types of populations. The alternative to Pool and Capie's null hypothesis is that the population average scores of pupils taught by trained and untrained teachers would be different.

As was true in the preceding example, since their experiment involved the comparison of only two population averages, Pool and Capie could have tested their null hypothesis by using a two-sample t-test. Instead they chose to use an equivalent analysis of variance.

In their footnote to Table II, the authors note that the F-statistic (3.347) is significant at the .065 level of probability. This means that the null hypothesis would have been rejected, had alpha been set at .065 or any larger value. Viewed another way, the sample averages for the treatment group and the control group (17.418 and 16.81, respectively; see Table I) are significantly different at the .10 level, but not at the .05 level. Had alpha been set at .05, the null hypothesis of equal population averages would have been retained.

Our Interpretation. As we have already noted, we feel that Pool and Capie were appropriately cautious in their interpretation of the results of their experiment. In contrast to the Get Set evaluation study discussed above, this research study is a true experiment. Subjects have been assigned to control and treatment conditions randomly, thus controlling the possibility that factors other than the experimental treatment were responsible for the significant difference between the sample average.

We agree with the authors' conclusion that pupils who were taught by the preservice teachers who received training in how to explain demonstrated better average performance on the concept attainment test. The appropriate generalization of this finding is another question, which we discuss below.

Unfortunately, Pool and Capie do not describe the composition of their samples of elementary school pupils. They imply that the samples included pupils in grades one through five, but give no indication of the distribution of pupils across these grades. Nor do they describe the samples in terms of mathematical aptitude, socioeconomic status, race, or any other background variables that might well influence performance on a test of attainment of a mathematical concept. So the limits of generalization to populations of elementary school pupils are difficult to determine.

The authors state that the preservice teachers used in their experiment were all in their junior year of a competency-based teacher training program at the University of Georgia. But again, the limits of generalization are uncertain becaue the sample of preservice teachers is not described in terms of age distribution, ability level, gender distribution, or major area of emphasis within education. Each of these variables might influence the effectiveness of the "How to Explain" instruction, and might therefore have an effect on generalizability.

Another consideration in the interpretation of the Pool and Capie results, which they note in passing, is their use of a specially constructed mathematical concept that was unfamiliar to the teachers as well as the pupils. This design feature eliminated a potential source of contamination in the experiment, since it ruled out the possibility that some teachers or pupils would know the concept prior to the experiment, while others would not. However, it limits the generalizability of the research findings since one cannot be sure that the "How to Explain" instructional materials would be as effective for teachers who were trying to explain familiar concepts.

Generalizing to Your Setting. We have discussed the potential generalizability of the Pool and Capie results at some length in the preceding section, and there is very little to add. If you were a school administrator considering the merits of using the Pool and Capie adaptation of the "How to Explain Program" with teachers in your school system, you would have to address several issues in addition to those we've already discussed. First, Pool and Capie's experiment involved preservice teachers. You would probably want to use the instructional materials with in-service teachers. Would the materials be as effective with teachers who had extensive classroom experience? You might want to use the materials with teachers who taught in grades other than one through five, and you would certainly not want to limit your use of the materials to mathematics teachers. You have no way of knowing whether the materials would be effective in training teachers of pupils in higher grades, or teachers of subjects other than mathematics. The Pool and Capie results are promising, but additional research is clearly needed to support the use of the "How to Explain" materials in an in-service teacher education program. In fairness to the authors, it must be noted that they do not recommend such generalization of their findings.

Summary

We have discussed three research and evaluation studies that incorporated analysis of variance to test a null hypothesis of equal population averages. In each case, the alternative hypothesis was that the population averages were different.

Note that the inferential results in each study were summarized in the form of an analysis of variance table. All of the analysis of variance tables contained the same basic information, although the arrangement of the tables varied a bit. You will find the same basic structure and information in all analysis of variance tables, as you encounter other studies that use this form of analysis.

When you read reports of research that incorporate the analysis of variance, keep these points in mind. Remember also that the analysis of variance presumes the research has collected data from independent random samples selected from each population of interest.

Once again, we hope you can now read and understand research and evaluation studies with greater clarity, confidence, and appropriate wariness—believing what is true and rejecting the inevitable falsehoods that plague research and evaluation in every field.

Problems—Chapter 12

In 1972, Neymann and Pratt reported the results of a study on the effectiveness of alternative methods of teaching beginning statistics at the college level. They randomly selected 75 students from among those registering for beginning statistics courses in the psychology department of a major university, and assigned independently selected groups of 25 students to each of three teaching methods: Traditional Lecture, Self-Paced Programmed Text, and a Discovery-Laboratory Method.

The authors administered a common final examination to the three groups, and reported their results in the following analysis of variance table:

	Sum of Squares	df	Mean Squares	F
Between Groups	162	2	81.00	5.67*
Within Groups	1028	72	14.28	
Total	1190	74		

* signif. @ 0.05; .95F (2,72) = 3.13

12.1 What null hypothesis and what alternative hypothesis were the authors testing in this study?

12.2 Assuming that the authors acted in accordance with the results reported above, what decision did they reach on retention or rejection of the null hypothesis? Support your answer by citing relevant data.

12.3 Interpret the results of this study in your own words. What assumptions are necessary to the interpretation you have made?

12.4 Describe a research or evaluation study in your own context in which one-way analysis of variance (ANOVA) would be an appropriate method of data analysis.
   a. Define the research question you might want to investigate.
   b. Identify the population(s) and population parameters that are central to your research study.
   c. State a null hypothesis and an alternative hypothesis that are relevant to your study.
   d. What conclusions would you reach, were you to reject the null hypothesis?

13.

Two-Way Analysis of Variance

The Nature of the Inferences

All of the problems we discussed in Chapters 9 and 12 (Inferences Involving Averages and One-Way Analysis of Variance) were similar in one respect; each analysis involved *one* **independent variable** (such as type of curriculum or treatment) and *one* **dependent variable** (such as student achievement or teacher attitude). In this chapter we'll consider problems that are somewhat more complex, because they will involve *more than one independent variable*. The problems can therefore be more realistic and more interesting. But we will stick to a single dependent variable, so that the waters don't get too murky.

We introduced one-way analysis of variance in the last chapter by discussing the hypothetical problem of selecting the best of three metric education programs. We suggested that a school administrator should purchase the curriculum that would result in fourth-graders achieving the highest average score on a test of metric knowledge. We began with the null hypothesis that the three candidate curricula were equally effective; i.e., that the population of fourth-grade students in the school system would earn the same average score regardless of the curriculum selected. The alternative hypothesis was that at least one of the three curricula was more effective than the others.

Let's consider the problem of selecting a metric education curriculum a bit further. We'll first examine the instructional approaches used in the three curricula, and then consider the aptitudes and preferences of students who might be instructed

257

using the curricula. Suppose that the three candidate curricula differ markedly in the way they present information about the metric system. Curriculum G presents information in a predominantly raphic form, with lots of pictures, charts and graphs. Curriculum V is highly *verbal* in its presentation. It uses lengthy verbal descriptions, but very few pictures. Curriculum S presents material in highly *symbolic* form, with equations that emphasize the relationships between English and metric measures.

Now it is well known that students differ in their preferences for the form in which new information is presented, and in the efficiency with which they learn information that is communicated in various forms. Some students are highly visual learners. Their most efficient learning mode is through study of graphs and pictures. Other students learn very little from visual materials, and do best with simple prose explanations. They are verbal learners. Still others learn most efficiently by studying relationships among symbols. Equations convey a great deal of meaning to such students, and provide useful summaries of important relationships. These facts suggest the possibility that no one metric education curriculum may be best for all students. Curricula G, V and S might provide the most efficient learning mechanism, and the highest possible achievement, for different groups of students.

Using one-way analysis of variance, we were able to examine the possibility that one of the three curricula was more effective than the others for the entire population of fourth-grade students. Recall that our null hypothesis was just the converse: that all three curricula were equally effective. One-way ANOVA did not permit us to explore the possibility that one curriculum might be best for one type of student, while another curriculum might be best for others. Using **two-way analysis of variance,** we can still entertain the possibility that one of the three curricula is an overall winner, but in addition we can examine the more intriguing possibility that different curricula are best for different types of students. We can even consider the possibility that students' preferences for a particular type of instructional material (graphic, verbal, or symbolic) influence average achievement on the test of metric knowledge, regardless of the curriculum material used. In short, two-way ANOVA facilitates the examination of a far richer set of evaluation and research questions than does one-way ANOVA. That's not to say it is always better or more appropriate. Both one-way and two-way ANOVA are useful in particular evaluation and research settings.

The research questions that we can investigate using two-way analysis of variance might be easier to understand by considering the following set of figures. Each figure illustrates one of a number of possible relationships between the type of curriculum used, students' preferences for different types of instructional materials, and students' average performance on a test of metric knowledge. In each of the figures, average achievement on the knowledge test is shown on the vertical axis, and type of materials is indicated on the horizontal axis, and type of curriculum used is indicated in the body of the figure.

Figure 1 illustrates the possibility that the three curricula differ consistently in their overall effectiveness, regardless of students' preferences for various types of instructional materials. Since each type of student has the same overall average achievement, this figure also suggests that the preference variable has no effect on students' performance on the test of metric knowledge. Using statistical language, we would describe the results shown in Figure 1 by saying three things: First, there *is* a **main effect** due to

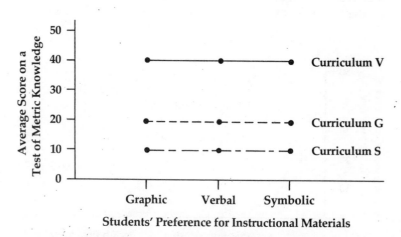

Figure 1. Illustration of Main Effect Due to Type of Curriculum, no Main Effect Due to Student's Preferences, and No Interactions.

type of curriculum used. Second, there is *no* main effect due to preference for type of instructional material. Third, there is *no* **interaction** between type of instructional materialand students' preferences. We'll discuss these terms in some detail when we've finished describing the figures. In the meantime, you should realize that we'd describe Figure 1 in exactly the same way, regardless of which curriculum won the achievement race. For example, if Curriculum S had resulted in the highest average achievement, and Curriculum V the lowest, we would still say that there was a main effect due to curriculum, and no other effects. The flat, parallel lines that connect the average achievements of each group of students are the key to this description.

Another possible experimental outcome is that the three *curricula* are equal in their *overall* effectiveness, that none of the curricula is more effective than any other if you ignore students' preferences for types of instructional materials, but that students differ in their average performance on the test of metric knowledge, depending on their preferences for types of instructional materials. Statistically speaking, with this outcome we would say that there is *no* main effect due to curriculum, that there *is* a main effect due to students' preferences for types of instructional materials, and that there is *no* interaction between curricula and students' preferences. This outcome is illustrated in Figure 2,

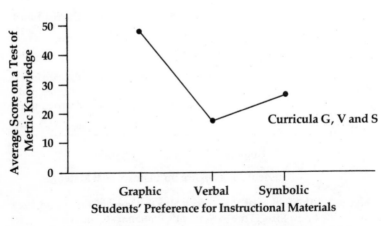

Figure 2. Illustration of No Main Effect Due to Curriculum, a Main Effect Due to Students' Preferences, and No Interactions.

below. Note that in Figure 2, the average score for students *with a given instructional preference* is the same, *regardless of the curriculum* used. However, students with a *preference for graphic instructional materials* have a higher average score than do students who prefer either verbal or symbolic materials. Again, we'd give the same statistical description of these results—a main effect due to students' preferences, no main effect due to curriculum, and no interactions—whichever group of students had the highest average achievement. The key here is that all three curricula resulted in identical average achievement for students with a given preference.

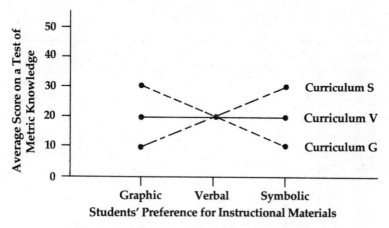

Figure 3. Illustration of No Main Effects Due to Curriculum or Students' Preferences, but an Interaction Between Curriculum and Students' Preferences

A third possible experimental outcome is illustrated in Figure 3. This figure illustrates the interesting possibility that no one curriculum is any more effective than any other when scores on the metric knowledge test are averaged across all fourth-graders in the population; that students with a preference for a particular type of instructional material have the same average score as students with a preference for any other type of material; but that *some combinations* of curricula and student preferences result in far higher achievement or far lower achievement than do other combinations. We would describe this outcome statistically by saying that there are *no* main effects due to curriculum or to student preferences, but that there *is* an interaction between type of curriculum and student preferences. Note that Curriculum G appears to be particularly effective when used with students who prefer that type of instructional material, that Curriculum S is particularly effective when used with students who prefer symbolic materials, and that type of curriculum appears to have no effect for students who prefer verbal instructional materials. This figure illustrates just one of many possible outcomes that would be described as illustrating "no main effects, but an interaction" between the independent variables. The labeling of the curriculum lines could be interchanged, for example, without altering the basic description.

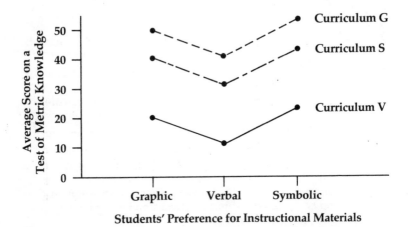

Figure 4. Illustration of Main Effects Due to Type of Curriculum and Students' Preferences for Instructional Materials, but No Interaction Effect

Each of the figures we've discussed so far illustrates one of the three effects that are possible in two-way ANOVA: a main effect due to one of the two independent variables, or an interaction effect between the independent variables. These effects can also be present in combination, and additional figures are needed to illustrate these possibilities. For example, a main effect due to each of the independent variables (type of curriculum and students' preference for instructional material) could be present, although there was no interaction between the independent variables. In this case, the three curricula would result in different average achievements for the entire population of fourth-grade students; the average achievement of students who preferred verbal materials would differ from the average achievement of students who preferred graphic or symbolic materials; but differences in the effectiveness of two curricula would be the same, for all three groups of students. Figure 4 illustrates this experimental outcome. Here again, we would describe the results shown in Figure 4 the same way—there are main effects due to type of curriculum and students' preferences, but no interaction—regardless of which curriculum was most effective or which group of students had the highest or lowest average achievement. The key to the description is that the lines that join the average achievements for a given curriculum are parallel to each other. Because of this parallelism, the difference in average achievement between students taught using Curricula V and S, or any other pair of curricula, is the same, for all groups of students. If the lines had not been parallel, specific combinations of curricula and student preferences would result in particularly high or particularly low average achievement, which we would describe as an interaction effect.

Three additional figures would exhaust the range of possible outcomes we have not yet illustrated. Two figures would show a main effect for one of the independent variables, in combination with an interaction effect. The third figure would show a "grand slam": a main effect due to each of the independent variables, in combination with an interaction effect. On the assumption that you've probably grown tired of looking at illustrative figures by now (regardless of your personal affinity for graphic instructional materials), and that you've grasped the general idea of how main effects and interactions make themselves evident in figures of this sort, we will skip the two intermediate possibilities, and end this section with an illustration of all three effects that can be present in two-way ANOVA. In Figure 5, the sets of average achievements are arranged to show a main effect due to type of curriculum; a main effect due to students' preference for instructional material; and an interaction between type of curriculum and students' preferences. Note that the lines that connect the average achievements associated with a given curriculum are not parallel in this case. In fact, when Curriculum G is used with students who prefer graphic materials, the resulting average achievement is very high—far higher than would result from mere addition of the effect due to use of Curriculum G with all students and the effect due to students who prefer graphic materials. In contrast, the use of Curriculum V with students who prefer symbolic materials results in exceptionally low average achievement. Here again, the combination of curriculum and students' preferences is the cause of the extraordinary achievement outcome. Neither independent variable would produce such results on its own, and the sum of the two effects (use of Curriculum V and preference for symbolic materials) is far exceeded by the effect of the combination.

In describing the results shown in Figure 5, we have implicitly defined an interaction effect. It is an effect due to a combination of independent variables that exceeds the sum of the main effects of the independent variables taken singly. An interaction effect is sometimes called a **synergistic effect**. In a homely but dramatic way, the combined effect of a lighted match and spilled gasoline illustrates **synergism** or interaction. Neither the presence of a match nor the presence of gasoline produces an explosion or fire. But in combination, the effect of the two variables is profound.

Two-way analysis of variance is an inferential procedure, as is one-way ANOVA. It makes use of data collected from several samples to test hypotheses about the parameters of the populations from which the samples were drawn. Three null hypotheses can be tested using two-way ANOVA. First, that one of the independent variables has no main effect; second, that the other independent variable has no main effect; and third, that there is no interaction effect between the independent variables.

Going back to our metric education example, let's discuss the applicable null hypotheses, and their implications. The first null hypotheses would be that there is no main effect due to type of curriculum. What we mean by this is as follows: Had all fourth-grade students in the school system been taught using Curriculum G, or Curriculum V or Curriculum S, their average score on the test of metric knowledge would have been exactly the same. In other words, type of curriculum has no influence on

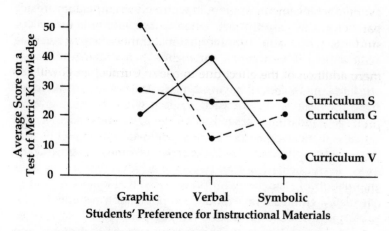

Figure 5. Illustration of Main Effects Due to Type of Curriculum and Students' Preferences and an Interaction Between Type of Curriculum and Students' Preferences for Type of Instructional Material

student achievement on the test of metric knowledge. The population parameter being considered here is the average test score of all fourth-grade students in the school system. Our null hypothesis is that this parameter is unaffected by the type of curriculum used; the average that would result from use of Curriculum G is equal to the average that would result from use of Curriculum V is equal to the average that would result from use of Curriculum S. The alternative hypothesis is that the average test score would be different, depending on the type of curriculum used. More specifically, if there is a main effect due to type of curriculum, at least one choice would result in a larger or smaller average score, for all fourth-graders in the school system.

The second null hypothesis that could be tested in this example is that there is no main effect due to students' preference for type of instructional material. Unfortunately, this statement is a little bit complicated, and our explanation will have to be correspondingly long-winded. Suppose that it were possible to conduct a survey of all fourth-graders in the school system, and that each student were to specify his or her preference for graphic, verbal or symbolic instructional materials. You could then divide the population of fourth-graders into three sub-populations, each identified with a preference. Suppose that you could instruct all fourth-graders with a preference for graphic materials, using each of the three curricula, administer the test of metric knowledge after you'd used each curriculum, and then wipe out the students' memories of having been instructed. If you averaged the three test scores for each student who preferred graphic materials, and then averaged all the averages, you'd have a grand average achievement for the sub-population of students who preferred graphic materials. In exactly the same way, it would be possible to calculate a grand average achievement for the sub-population of students who preferred verbal materials, and to calculate a grand average for the sub-population of students who preferred symbolic materials. At this point you'd have three sub-population averages. The null hypothesis that there is no main effect due to students' preference for type of instructional materials states that these sub-population averages are all equal. The alternative hypothesis is that one of the sub-populations has a higher average score on the test of metric knowledge than do the other sub-populations.

We have already discussed the third null hypothesis that can be tested using two-way ANOVA—the null hypothesis that there is no interaction effect between type of curriculum and student's preference for type of instructional materials. However, the concept of an interaction effect is novel, and it bears additional consideration. Perhaps the simplest way to describe the null hypothesis of no interaction effect is as follows: It states that if any sub-population were to be instructed using any one of the three curricula, the resulting average achievement could be calculated from the main effect due to that sub-population, and the main effect due to that curriculum. In other words, if there is no interaction effect, the main effects can merely be added together to determine the combined effect of the two independent variables. The alternative hypothesis, that there is an interaction effect, means that some combination of the two independent variables produces a result that is not equal to the sum of the main effects. In the case of the metric education program, for example, an interaction would be present if the sub-population of students who preferred graphic instructional materials, when instructed using Curriculum G, earned an average score on the test of metric knowledge that was higher than either their grand average, or the average of all sub-populations when instructed using Curriculum G.

You will find that our metric education example is a model for the inferences considered in the real studies that follow. In each study, the authors test null hypotheses on main effects of independent variables and on interactions between independent variables. In the second study, the authors considered three independent variables rather than two. They conducted what is called a **three-way analysis of variance.** We have included their study in this chapter to illustrate the generality of the principles introduced in discussing two-way ANOVA. Using those principles, it is possible to test null hypotheses on the main effects of and interactions among any number of independent variables. More null hypotheses are involved, and the interpretations become more complex, but the basic statistical approach is identical.

A Final Evaluation of a Career Education Program

Purpose of the Study. In Chapter 9 we discussed an evaluation of the third-year operation of a career education program by Kershner and Blair (1975). We'll return to that example here, to describe a component of the evaluation that made use of two-way analysis of variance.

Kershner and Blair conducted a summative evaluation of a career education program that was developed and operated by Research for Better Schools, Inc. The program was designed to provide secondary school students with "cognitive skills, career experiences, and personal perspectives which aid in the selection and pursuit of adult life goals" (p. 6). The program focussed on three areas of instruction: career exploration and specialization, career guidance, and basic skills. Many of the instructional activities took place in a large urban high school, and others took place at various community resource sites. Most of the program participants were eleventh-graders, but students from grades nine, ten and twelve were also served.

The evaluation had several purposes. Among them was the goal of comparing the attitudes of program participants to those of students in control groups, toward various aspects of career planning and knowledge. In this example we will limit our attention to this goal.

Evaluation Design. The evaluation involved assessments of the pre-treatment and post-treatment status of students in three experimental groups, and parallel assessments of the status of students in two comparison groups. Instruments used by the evaluators included measures of career maturity, students' self-acceptance, school attitudes, achievement, and career attitudes.

One of the experimental groups (labeled E1) and one of the control groups (labeled C1) were randomly selected for evaluation of the program during its third year of operation. Another experimental group (labeled E2) and another control group (labeled C2) were not selected randomly, but participated in the program during two years of its operation. The third experimental group (labeled E3) consisted of students who had been selected during the third year of program operation to replace those who had dropped out of the evaluation. The results we will consider in this example involve only the randomly-selected experimental and control groups (E1 and C1).

Kershner and Blair administered a fifteen-item Attitude Toward Career Knowledge Scale (ATCK) to all students in groups E1 and C1 at the beginning and the end of the third year of the program. Their purposes were to compare the attitudes of experimental and control-group students both prior to treatment and at the end of one year of program participation. Their results are shown below.

Authors' Data Table. Table 30 in the Kershner and Blair report (1975; p. 67) contains the results of two-way analyses of variance using the pretest and posttest performance of experimental and control-group students on the Attitude Toward Career Knowledge Scale. It is reproduced below.

Authors' Interpretation. Kershner and Blair (1975; p. 67) interpret the results shown in their Table 30 as follows: "The analysis of variance of the ATCK pretest scores indicated no between group difference. There was a significant grade difference which reflected a possible grade linked development of career maturity. The posttest analysis of variance indicated significant grade and group differences; E1 posttest performance on the ATCK was significantly higher than C1 performance. The grade differences were again identified and reinforced the stratification used in the analyses."

Attitude Toward Career Knowledge
E1 and C1 Groups
(Authors' Table 30)
Analyses of Variance

Source	df	MS	F
PRETEST			
Grade	3	108.2237	3.9814
Group	1	10.0928	0.3713
Grade x Group	3	14.7952	0.5443
Error	184	27.1823	
POSTTEST			
Grade	3	222.4236	7.9037
Group	1	164.7753	5.8552
Grade x Group	3	46.6252	1.6568
Error	184	28.1417	

Critical value F (1,180; alpha = .10) $\geq$ 2.74
Critical value F (3,180; alpha = .10) $\geq$ 2.12

The Nature of the Inferences. The data in Table 30 illustrate the results of testing six different null hypotheses. Three of the hypothesis tests involve pretest data and the other three involve posttest data. In each case, the authors are treating their experimental and control groups as though they had been randomly sampled from populations of students who might have participated in the career education program.

In another portion of Table 30, which we have not reproduced here, Kershner and Blair show separate average pretest and posttest scores on the ATCK for experimental and control-group students sampled from grades 9, 10, 11, and 12. They also show average pretest and posttest scores for all students in the experimental and control groups, regardless of grade level, and average scores for students at each grade level, regardless of group membership.

From the data in Table 30, we can tell that the authors were interested in analyzing the effect of two independent variables—grade level and treatment—on average ATCK scores. They examined not only the main effects of these two independent variables, but a possible interaction effect as well. The two portions of

Table 30 reproduced above show the results of separate but parallel analyses using pretest scores and posttest scores as dependent variables.

The first null hypothesis tested was that grade level had no main effect on ATCK pretest scores. This hypothesis reflects the following *hypothetical* scenario: Suppose that the authors could take an entire population of ninth-graders who were eligible to participate in the career education program, and divide it in half. One half of the population would be randomly selected for participation in the career education program, and the other half would be selected for the control group. Before any students had participated in the program, the authors would administer the ATCK to every ninth-grader in the population and record their average score. The authors would then complete the same procedure using tenth-graders, eleventh-graders and twelfth-graders. The null hypothesis states that the average pretest ATCK scores of the ninth-graders, tenth-graders, eleventh-graders, and twelfth-graders would be exactly the same. That is the precise meaning of the statement that "grade level has no main effect on ATCK pretest score." The alternative hypothesis would be that grade level *does* have a main effect on pretest score. This is equivalent to stating that the average pretest score of students in one grade will be different from the average pretest score of students in at least one other grade.

Kershner and Blair rejected the null hypothesis that grade level had no main effect on pretest score. They took this action because the F-statistic associated with the source labeled "Grade" in the pretest analysis of variance table (the number 3.9814) was larger than the critical value for an alpha level of 0.10. The appropriate critical value (2.12) is shown in the footnote at the bottom of Table 30. The degrees of freedom shown for this critical value of the F distribution (3, 180) were found from the value 3 in the df column for Grade, and the value 184 in the df column for Error. (Tables of the F distribution don't list critical values for every choice of degrees of freedom. Often a researcher must use the critical value associated with degrees of freedom that are close to the desired value. Using the next smaller value for degrees of freedom in the denominator is a conservative decision, so Kershner and Blair appropriately used 180 instead of 184). In case you've forgotten what degrees of freedom and F-statistics are, you can refresh your memory by looking back to Chapter 12, on One-Way Analysis of Variance.

The second null hypothesis that Kershner and Blair tested was that their "treatment" had no main effect on pretest score. This null hypothesis might seem strange to you, since at the time the pretest was administered, none of the students had received a "treatment"; neither the students in the experimental group nor the students in the control group had participated in the career education program at that point. The function of this hypothesis test was to provide statistical assurance that the random assignment of students to the experimental and control groups had been effective. Had the null hypothesis been rejected, the authors would have concluded that the experimental and control samples were chosen from populations of students that differed in their average scores on the Attitude Toward Career Knowledge Scale, at the beginning of the school year. Fortunately, the null hypothesis that the populations of potential experimental and control students had identical average scores on the ATCK at time of pretest was retained. We can substantiate the appropriateness of the authors' conclusion by noting that the F-statistic for Group (0.3713) is much smaller than the critical value (2.74) for an alpha of .10 shown in the footnote to Table 30. This critical value is for an F distribution with 1 degree of freedom in the numerator (the value shown in the "df" column for Group) and 180 degrees of freedom in the denominator (a value that is just smaller than the 184 shown in the "df" column for Error).

The final null hypothesis tested for the pretest dependent variable was that grade level and treatment had no interaction effect. The F-statistic associated with this test is shown in the Grade x Group row of the Pretest section of Table 30. Once again, the F-statistic (.5443) is much smaller than the associated critical value (2.12) corresponding to an alpha level of 0.10, so the null hypothesis of no interaction effect was retained by the authors. Notice that the "Grade x Group" source has three degrees of freedom associated with it. It is no accident that the degrees of freedom associated with this interaction is the product of the df values associated with the independent variables (Grade and Group) that form the interaction (in this case $3 = 3 \times 1$). Degrees of freedom associated with an interaction term will always equal the product of the degrees of freedom associated with the sources that form the interaction. The critical value 2.12 was appropriate here because the interaction term Grade x Group had 3 degrees of freedom and the Error source had 184 degrees of freedom associated with it.

Had the Grade x Group interaction been statistically significant, it would have meant that the random assignment of students to experimental and control groups was flawed in at least one grade. This is true because the alternative hypothesis in this case is that the difference between average pretest ATCK scores for potential experimental-group and control-group students in one grade was not equal to the corresponding difference for students in at least one other grade. Since there was no main effect due to Group, a significant interaction effect could only be interpreted as a significant control-group versus experimental-group difference in at least one grade.

The section of Table 30 headed "Posttest" contains the results of three hypothesis tests that parallel those described above for the pretest dependent variable. The row labeled "Grade" contains the results of a test of the null hypothesis that grade level had no main effect on average posttest score. As the authors state, this null hypothesis was rejected. The F-statistic shown for Grade (7.9037) is substantially larger than the critical value (2.12) associated with an alpha level of 0.10 for an F distribution with 3 degrees of freedom in the numerator and 180 degrees of freedom in the denominator.

In rejecting the null hypothesis that grade level has no main effect on posttest score, the authors are stating that the average ATCK scores of the populations of potential program participants from which the samples of ninth-graders, tenth-graders, eleventh-graders and twelfth-graders were drawn are not equal. At least two of these populations have different average scores.

The second null hypothesis tested in this section was that participation in the career education program had no effect on average posttest ATCK scores. As the authors report, this null hypothesis was also rejected. Since the null hypothesis concerning a main effect due to treatment was retained for the pretest dependent variable, the authors are on reasonably firm ground in concluding that the career education program was effective in influencing students' attitudes toward career knowledge.

The relevant statistics for the main effect of treatment (Group) on average posttest scores are the F value of 5.8552 and the critical value for an alpha level of 0.10 and an F distribution with one degree of freedom in the numerator and 180 degrees of freedom in the denominator (2.74). Since the F-statistic exceeded the critical value, the null hypothesis of no treatment effect was rejected.

The final null hypothesis tested by Kershner and Blair was that grade level and treatment had no interaction effect on average posttest score. This null hypothesis was retained, since the associated F-statistic (1.6568 shown for the Grade x Group source) was smaller than the critical value of 2.12 shown in the footnote to Table 30.

Had the null hypothesis of no interaction effect been rejected, the authors would have concluded that their treatment was more effective for students in one grade than for students in at least one other grade. In fact, they might have found that the career education program influenced the average ATCK scores of experimental-group students in one or more grades, but had no such influence for students in the remaining grades. In statistical terms, a significant interaction effect would have meant that the difference between the average posttest ATCK scores of the populations of potential experimental and control-group students varied across grade levels. For example, the difference between experimental and control-population averages might have been larger for tenth-graders than for twelfth-graders.

Our Interpretation of Results. We agree fully with the authors' conclusion that the career education program had a statistically significant influence on the participating students' attitudes toward career knowledge, as measured by the ATCK. Whether or not this influence is educationally or substantively significant is another question. This latter question could be examined more adequately by computing a confidence interval on the difference between the average posttest ATCK scores of the experimental and control populations. Upon examining this confidence interval, you would have to decide whether the difference in average scores was "large enough to be important." As is always the case, statistics can facilitate a judgment on the question of substantive significance, but do not provide an answer to that question. In the final analysis, statistics is only part of a complete evaluation; you will have to look at the needs assessment for the answer here.

Assumptions Needed to Generalize. Were you to use these experimental results as a basis for adopting the career education program in your school system, you would be making a number of important assumptions. These assumptions were described at length in Chapter 9, when we last considered the Kershner and Blair (1975) study. You might want to review that discussion now.

An Evaluation of a Multicultural Education Program

Purpose of the Study. Alexander and Brown (1978) conducted a unique evaluation of a multicultural education program operated by the Houston, Texas, school system. The purpose of the program was to provide elementary school students with diverse cultural experiences, in order to increase their respect for persons of other cultures and to promote their understanding and acceptance of persons of other races. We'll discuss the statistical and testing aspects of the evaluation.

The program exposed students to the cultures of six different nations or continents—Africa, China, Germany, India, Mexico, and the United States. Each culture was presented in a separate classroom, and involved students in the geography, attire, art, music, dance, lifestyle, language, food, games, and crafts of the region represented.

Students who participated in the program were selected from grades two through six. Equal numbers of black, Anglo, and Mexican-American students participated in the program.

The purpose of the evaluation was to determine whether program participation led to changes in students' attitudes toward children of other racial-ethnic groups. In particular, the evaluators tried to determine whether program participants became more accepting of students of other racial and ethnic backgrounds.

Evaluation Design. Alexander and Brown used a simulation game to assess the pre-treatment and post-treatment attitudes of students in an experimental group (program participants) and a control group. The game consisted of having students place three six-inch dolls in a scale-model classroom. One of the dolls represented a white child, one represented a black child, and one represented a Mexican-American child. The dolls were distinguished by color and other physical features, and the evaluators confirmed that the students made the correct racial-ethnic classifications. Each student in the experimental group was told to identify one of the dolls as him/herself. The student was then told that his/her doll had to wait in the model classroom for the instructor to come, and that the two children represented by the other dolls also had to wait. The student was instructed to place the doll representing him/herself anywhere in the model classroom and then to place the dolls representing the other two children anywhere in the classroom. Each student in the control group was given the same instructions, except that these students were told that they had to wait for a bus driver to take them back to school. Each student participated in the exercise independently.

When a student had finished placing the dolls, their positions were noted on a piece of paper that formed the floor of the model classroom. The total distance between the dolls was used as a measure of proximity.

The doll placement exercises were conducted twice: once prior to the experimental students' participation in the multicultural education program, or the control-group students' participation in a school function of similar duration that did not feature culturally-related experiences or instruction, and once immediately following such participation.

There were 270 program participants, randomly selected from students in grades two through six. Ninety of the participants were white, 90 were black, and 90 were Mexican-American. The control sample consisted of 90 students sampled in the same proportions from each racial-ethnic group and each of grades two through six. The authors do not indicate the population from which participants were sampled, nor do they state that control-group students were randomly sampled.

Alexander and Brown assumed that students who placed dolls of racial or ethnic groups other than their own closer to their own doll would be more accepting of students of other races and cultures, than would students who placed the other dolls farther from their own. They thus used proximity of doll placement as a measure of racial and cultural acceptance.

The dependent variable used in the Alexander and Brown evaluation was the total distance between the dolls in the pre-treatment measure, minus the total distance between the dolls in the post-treatment measure. The authors reasoned that a change in students' acceptance of children of other races and cultures would be indicated by closer placement of the dolls in the post measure than in the pre measure. A positive pre-minus-post difference in total distance would therefore correspond to a positive change in student attitude.

Authors' Data Table. Alexander and Brown conducted a three-way analysis of variance, using the pre-minus-post difference in total distance scores as a dependent variable. Their independent variables were grade level (grades two, three, four, five and six), racial-ethnic group membership of students (white, black, and Mexican-American), and instructional experience (multicultural education or control-group participation). Table 1 in the Alexander and Brown paper (1978; p. 8) contains the results of their analysis. It is reproduced below, just as it appeared in the paper.

Authors' Interpretation. Alexander and Brown concluded, on the basis of the results shown in their Table 1, that type of instructional experience had a significant main effect, that racial-ethnic group had a significant main effect, and that grade level of

(Authors' Table 1)
Analysis of Variance Summary Table

Source	SS	DF	MS	F
Ethnicity	732.38	2	366.91	3.66*
Grade	1056.49	4	264.13	2.64*
Instructional Experience	4784.87	1	4784.87	47.83*
Ethnicity by Grade	680.78	8	85.04	.85
Ethnicity by Instruction	38.91	2	19.46	.19
Grade by Instruction	1478.66	4	369.66	3.69*
Ethnicity by Grade by Instruction	1020.96	8	127.60	1.27
Within	33011.39	330	100.04	

*p < .05

student had a significant main effect. They reported that, on average, students who participated in the multicultural educational experience placed the dolls eight centimeters closer together in the posttest than they had in the pretest. In contrast, students who were members of the control group placed the dolls only an average of 0.59 centimeter closer together in the posttest than in the pretest. The authors stated that the scores of "students who participated in the [multicultural education experience] were significantly different from the scores of the students participating in the neutral school function [control-group students]." (1978; p. 7)

The Nature of the Inferences. The authors' Table 1 summarizes the results of testing seven different null hypotheses. We will discuss them in order.

Alexander and Brown first tested the null hypothesis that students' racial-ethnic group had no effect on the average value of the dependent variable (change in the proximity of the dolls). To understand the meaning of this hypothesis, it is again necessary to construct a hypothetical scenario. Suppose that the students assigned to the experimental group and the control group had been sampled from three populations that differed only in their racial-ethnic background; a population of white second-through sixth-graders, a population of black second- through sixth-graders, and a population of Mexican-American second-through sixth-graders. Suppose further that each population had been divided in random halves. One half of each population had completed the doll placement exercise, then had participated in the multicultural education program, and then had completed the doll placement exercise again. The other half of each population had completed the doll placement exercise, then had participated in the control-group experience, and then had completed the doll placement exercise again. Finally, suppose that the evaluators had computed the average value of the dependent variable for each population. There would be one average change in doll placement distances for students in the white population, one average change for students in the black population, and one average change for students in the Mexican-American population. The null hypothesis states that these average changes would all be equal. The alternative hypothesis is that at least one of these average changes would differ from another.

The authors rejected the null hypothesis that students' racial ethnic group had no main effect on the dependent variable. In doing so, they compared the F-statistic for Ethnicity (3.66) with the critical value of an F distribution with two degrees of freedom in the numerator and 330 degrees of freedom in the denominator, corresponding to an alpha level of 0.05. Since 3.66 was larger than the critical value, the authors rejected the null hypothesis at the 0.05 level of significance. We can tell that Alexander and Brown reached this decision because the 3.66 value in Table 1 has a "*" beside it, and the footnote to Table 1 explains that the "*" indicates a probability value (alpha level) that is less than .05.

In a similar fashion, Alexander and Brown tested, and rejected, null hypotheses that grade level of students and instructional experience had no main effects on the dependent variable. Their null hypothesis concerning grade level applies to populations of students from grades two, three, four, five, and six who might have participated in either the multicultural education program or the control-group experience. Interpretation of this null hypothesis requires a hypothetical scenario that is similar to the one we've just presented, in which half of each population was randomly assigned to the multicultural education program and the other half was assigned to the control-group experience. The null hypothesis states that the average score on the dependent variable would be exactly the same, for students at each grade level. The alternative hypothesis states that the average scores would be different, for at least two of the populations.

The null hypothesis that instructional experience has no main effect on the dependent variable is the one most important to the evaluation. This null hypothesis presumes that, if populations of students from each grade level and each ethnic group had been divided in half randomly, and if one half of each population had been assigned to the multicultural education program and the other half to the control-group experience, the two halves of the populations would have exactly the same average score on the dependent variable. The alternative hypothesis is that the average score of students who had been assigned to the multicultural program would differ from the average score of students who had been assigned to the control group.

Since the F-statistic associated with instructional experience (47.83) is larger than the alpha = .05 critical value of the F distribution for one and 330 degrees of freedom, the difference between the sample average values on the dependent variable was declared to be significant at the .05 level; that is, the null hypothesis was rejected.

Alexander and Brown tested three null hypotheses concerning interactions between pairs of independent variables. The first was that students' racial-ethnic group and grade level would not interact in their effect on the dependent variable. The meaning of this null hypothesis is as follows: Suppose that populations of black, Anglo and Mexican-American students had different average scores on the dependent variable. According to the null hypothesis, the differences between their average scores would be the same, regardless of the students' grade levels. This null hypothesis was sustained, because the associated F-statistic (.85) did not exceed the critical value of an appropriate F distribution for an alpha level of .05.

The second null hypothesis concerning an interaction stated that students' racial-ethnic group and instructional experience would not interact in their effect on the dependent variable. To understand the meaning of this null hypothesis, assume once again that the authors had three populations of students (one of each racial-ethnic group represented in the study) and that half of each population had been randomly assigned to each instructional experience. Suppose that the halves of each population that were assigned to the two instructional experiences had different average scores on the dependent variable. The null hypothesis states that these differences would be exactly the same, for the black, Anglo, and Mexican-American populations. Since the F-statistic associated with this null hypothesis (.19) did not exceed its corresponding critical value, the null hypothesis was retained.

The third null hypothesis involving an interaction between two independent variables was rejected. The results in Table 1 show that there was a significant grade level by instructional experience interaction, since the associated F-statistic (3.69) has an "*" beside it.

The null hypothesis that there was no grade level by instructional experience interaction means the following. Suppose populations of students in each of grades two through six had been divided randomly in half, with one half assigned to the multicultural education program and the other half assigned to the control-group experience. The null hypothesis states that, for students in each grade, the two halves of the population would have average scores on the dependent variable that differed by exactly the same amount. In other words, if the null hypothesis were true, it would mean that the difference between the effectiveness of the experimental and control-group treatments would be exactly the same for second-graders as it was for third-graders, and fourth- through sixth-graders.

Since this null hypothesis was rejected, we can surmise that the experimental treatment resulted in a greater change in doll placement distance for students in some grades than for students in others.

The final null hypothesis tested in this study involved what is called a **three-way interaction.** The null hypothesis states that there is no interaction effect of students' racial-ethnic group membership, grade level, and the instructional experience to which they were assigned. Three-way interactions are very complicated, and they can be described in a variety of ways. Perhaps the simplest explanation in this study is as follows: The data in Table 1 show that there was an interaction between students' grade level and their instructional experience. The null hypothesis that there is no three-way interaction states that the interaction between students' grade level and instructional experience is exactly the same, for populations of students in each racial-ethnic group. This is an admittedly terse explanation. Because a full explanation would require many additional pages, we will take the easy path of referring you to a traditional statistics text (Glass and Stanley, 1970) for more information. The null hypothesis that there was a three-way interaction between the independent variables was retained, since the associated F-statistic (1.27) was smaller than the alpha = .05 critical value in an appropriate F distribution.

Well, we've finally considered all seven null hypotheses represented by the results shown in Table 1. The authors chose not to interpret any of the hypotheses involving interactions, since they were concerned primarily with main effects and only one null hypothesis involving an interaction was rejected. This decision is, perhaps, warranted in this study. However, *interaction effects can often render main effects uninterpretable,* so it is generally foolhardy to ignore interactions. Although it is highly unlikely in this study, it is possible that students in one grade who were assigned to the control-group experience had greater average change on the dependent variable than did students who were assigned to the multicultural education program. For students in other grades, just the reverse might be true. This could happen even though there was a main effect due to instructional experience. To fully understand the meaning of main effects and interactions, it is often necessary to construct graphs similar to those shown in Figures 1 through 5 at the beginning of this chapter.

You may have wondered about the row in Table 1 that is labeled "Within." The data in this row of the table represent the variation in the dependent variable within each of the groups in the experiment. In some studies, this row of the ANOVA table is labeled **error** or **residual,** instead of "Within." The terms are synonymous in this application.

Our Interpretation of Results. In many ways, Alexander and Brown did an excellent job of discussing the implications of their findings. They are particularly sensitive to questions concerning the validity of their doll placement exercise as a measure of racial attitudes.

As in all evaluative research, it is necessary to go beyond the statistical information when judging the educational implications of the results. In the case of this study, we must be concerned, as were Alexander and Brown, with the meaning of a main effect due to instructional experience. Also, because it is not certain that the experimental and control groups were randomly sampled from the same populations, we must be cautious in stating that the multicultural education program alone was responsible for the main effect. Finally, the populations from which the samples were selected are not well described, so we are uncertain about the generalizability of the results of this study. Perhaps most important, we have no idea how much of a change in student attitudes is represented by the change in doll placement distances the authors found.

The measurement approach used in this study is unusual and imaginative. Measuring attitudes is very difficult, and it is doubly difficult when young children are involved. Nevertheless, it is the responsibility of the evaluators to demonstrate that the changes they found represent real and important differences in students' attitudes toward children of other races and cultures.

Assumptions Needed to Generalize. Were you to use the results of this study as a basis for adopting the multicultural education program in your setting, you would have to make several strong assumptions. First, you would have to assume that it was important to change students' attitudes toward children of other races and cultures. Second, you would have to assume that the doll placement exercise was a valid measure of students' attitudes. Third, you would have to assume that the changes observed in this study represented important modifications of students' attitudes. Finally, you would have to assume that the program was transportable; i.e., that the results found in this study would be likely to occur in your own setting. These are all strong assumptions. We would therefore encourage you to be appropriately cautious about any generalizations to your own situation—even without looking at costs, alternatives to, and side-effects of the program.

Summary

Although this chapter is long, we introduced relatively few new concepts. You've probably realized that the statistical ideas we introduced in Chapter 12 were used heavily in this chapter as well. The basic logic of analysis of variance applies here as it did in the last chapter. Both one-way and two-way ANOVA make use of sums of squares, degrees of freedom, mean squares, F-statistics, and the F distribution to test null hypotheses. The same kinds of statistics are also used in three-way ANOVA.

The most important new idea introduced in this chapter was the notion of an interaction effect. Whenever evaluators and researchers investigate the effects of more than one independent variable at a time, they must be sensitive to the possibility that those variables may interact. Often a single variable will have different effects under different conditions.

The main advantage of two-way and higher-way analysis of variance is that these statistical procedures allow researchers and evaluators to examine more-complex relationships than do the techniques we discussed in earlier chapters. Nothing is simple in education or the social and behavioral sciences. Effects rarely have a single cause. Treatments that are effective for one group of students might be ineffective for other groups. In order to discover what works and what doesn't, and to learn how things work, we need statistical procedures that allow us to model nature in its true complexity. Two-way analysis of variance is a small first step in the right direction.

Problems—Chapter 13

In 1974, Neymann and Pratt expanded and replicated their study on the relative effectiveness of various methods of teaching beginning statistics at the college level (See Problems for Chapter 12). In this replication they investigated the relative effectiveness of the three methods (Traditional Lecture, Self-Paced Programmed Text, and a Discovery-Laboratory Method) for students at two levels of quantitative aptitude (Low Aptitude and High Aptitude).

The researchers randomly selected 90 students from among those who registered for a beginning statistics course in the Department of Psychology of a major university. They administered a standardized quantitative aptitude test to the 90 sampled students, and divided the group at their median score into a high aptitude group (those who scored above the median) and a low aptitude group (those who scored below the median). Neymann and Pratt then randomly assigned one-third of the low aptitude group and one-third of the high aptitude group to each of the three instructional methods. At the end of a one-semester instructional period, the researchers administered a common final examination to all 90 students.

The results of the study were reported by Neymann and Pratt in the following analysis of variance table:

Source	Sum of Squares	df	Mean Squares	F
Instructional Method	190.62	2	95.31	5.40**
Aptitude level	161.85	1	161.85	9.17**
Method × Aptitude	128.84	2	64.42	3.65*
Within Groups	1482.60	84	17.65	
Total	1963.91	89		

 * signif. @ 0.05 level   **signif. @ 0.01 level

13.1 The authors tested three null hypotheses in this study; two were concerned with main effects and one with an interaction effect. Describe these null hypotheses.

13.2 Assuming that the authors acted in accordance with the results reported above, what decisions did they reach on retaining or rejecting the three null hypotheses? Support your conclusions by citing relevant data from the table.

13.3 Interpret the results of this study in your own words. What assumptions are necessary to the interpretation you have made?

13.4 In light of the significant interaction effect noted in the data table, can you safely conclude that one of the instructional methods is more effective than the others for all students, regardless of their quantitative aptitude level? Discuss this point.

14.

Some Advanced Topics

Introduction

In this chapter we introduce three advanced statistical procedures that are used quite frequently in evaluation and research studies. The first, called the **analysis of covariance,** is an extension of the analysis of variance. The second is **multiple linear regression analysis.** This procedure is particularly useful in studies that involve prediction (such as prediction of students' achievement from their aptitude scores and data on the extensiveness of their instruction). The third procedure is called **factor analysis.** It is often used when constructing attitude scales or tests. Factor analysis allows a researcher to create a few variables (called factors) that do a good job of representing a lot of variables. It is therefore used to make things simpler—an always laudatory goal!

In the balance of this chapter we'll describe each statistical procedure and then discuss one actual research or evaluation study that makes use of each procedure. Discussion of the analysis of covariance often takes up a healthy chunk of an advanced statistics course, so our coverage here will be relatively superficial. Likewise, many universities devote entire graduate courses to regression analysis and factor analysis, so we won't be telling you all there is to know about either of these topics in this relatively brief chapter. Nonetheless, we feel it is important that you gain some sense of the purpose and power of these advanced statistical methods so that they won't be totally foreign when you encounter them in evaluation and research reports.

Analysis of Covariance

The analysis of covariance is conceptually similar to the analysis of variance procedures you've already encountered in Chapter 12 and 13. Its basic purpose is to allow a researcher or evaluator to test the null hypothesis that two or more population averages are equal, against the alternative hypothesis that at least one of the population averages is different from the others. For example, using analysis of covariance procedures you could test the null hypothesis that a population of sixth-grade students would have the same average score on a test of functional writing skills, whether the students were taught to write in standard English classes (Curriculum A), in creative writing courses (Curriculum B) or in special writing drill-and-practice sessions (Curriculum C). In terms of the fundamental research and evaluation questions that can be addressed, analysis of covariance is just like the analysis of variance.

The analysis of variance and the analysis of covariance differ in two basic ways. First, the analysis of covariance allows a researcher or evaluator to address additional questions that the analysis of variance does not consider. Second, the two procedures differ in their detailed statistical computations. Consider once again the example involving instruction in functional writing. Were you to use the analysis of variance to test the null hypothesis of equal average performances on the functional writing test, you would select a simple random sample of sixth-graders from the population of interest, and then divide that sample randomly into thirds. One-third of the students would be assigned to Curriculum A, another third to Curriculum B, and the final third to Curriculum C. At the end of the school year, you would administer the functional writing test to all three samples of students, and use their average post-treatment scores to decide whether or not the null hypothesis of equal population averages was tenable. If the average scores of the three samples of students differed sufficiently, you would reject the null hypothesis that the three population averages were equal.

In using analysis of variance as we have described it, you would assume that the process of randomly dividing the sample of sixth-grade students into three parts had been effective in making the three resulting samples of students equivalent in all ways that were relevant to their writing test performance. In other words, you would assume that the students assigned to each of the three curricula were equally capable at the outset, and that random assignment had ensured this equality. On the average, the assumption is a good one, particularly if the three samples of students are reasonably large. However, even though random sampling and assignment ensures that samples are equivalent on the average, it does not ensure that *every* random sample is equivalent to every other one. Random assignment is particularly vulnerable to the vagaries of Lady Luck when sample sizes are small. It might not be sufficient for you to know that the samples of students assigned to Curricula A, B, and C will be equivalent in their capabilities *on the average,* since you will conduct your experiment only once. It would be far better to have some assurance that the particular samples constructed for your experiment were equivalent, and to correct for differences between the samples, should they exist. This is where the **analysis of covariance** can help.

Suppose you believed that reading skill was an important predictor of a student's functional writing ability, and that you had scores on a standardized reading test that had been administered to all sixth-grade students, just prior to their assignment to the three curricula. Using the analysis of covariance, you could then: 1) examine your belief that performance on the reading achievement test was a predictor of functional writing achievement following instruction; 2) examine the hypothesis that the relationships between students' scores on the reading achievement test and their scores on the functional writing test were the same, regardless of the curriculum used to instruct them; and 3) adjust for differences in the average pre-instruction reading achievement of the three samples of students, should such differences be found. In short, the analysis of covariance would allow you to correct for some important differences between the three samples that might still be present, even though the samples had been constituted through a random assignment procedure. By making the samples more "alike" at the outset of the experiment, you would add strength to an assertion that any differences in average functional writing scores at the end of the experiment were due solely to differences in the effectiveness of the three curricula.

Three variables were involved in the hypothetical experiment we just described. Type of curriculum (A, B, or C) is an **indepen-**

dent variable. It is the treatment variable whose effect is to be ascertained. Reading achievement score prior to instruction is another independent variable that, in this experiment, plays the role of a **covariate.** A covariate is a variable that is correlated with the **dependent variable.** It is measured prior to the administration of a treatment, and might account for some portion of the differences between the average scores of the treatment groups on the dependent variable. Functional writing test score at.the end of the experiment is an example of a dependent variable. In the analysis of covariance, we attempt to determine the effect of the independent variable on the average value of the dependent variable, after between-sample differences on the covariate have been eliminated. In the experiment we've just described, we would attempt to determine the effect of type of curriculum on average functional writing scores, after we had eliminated the effect of differences between the average initial reading achievement scores of the experimental samples.

The hypotheses that can be examined using the analysis of covariance are illustrated in Figures 1 and 2, shown below. In each figure, scores on the dependent variable are shown on the vertical axis, and scores on the covariate are shown on the horizontal axis. For the functional writing experiment, students' scores on the writing test would be plotted on the vertical axis, and their scores on the reading test that was administered prior to the writing instruction would be plotted on the horizontal axis. In Figure 1, the three diagonal lines represent the relationships between scores on the reading achievement test, and average scores on the writing test, for the samples of students taught using Curricula A, B, and C. The egg-shaped patterns (ellipses) around the three diagonal lines represent the distributions of scores on the reading test and the functional writing test for the three groups of students. For each group, average scores on the reading test and the functional writing test would fall at the center of that group's ellipse.

We have drawn Figure 1 as though there was a definite relationship between students' scores on the reading test, and their scores on the writing test administered after instruction. In fact, the figure suggests that there is an increasing relationship between the two variables (the higher the score on the reading test, the higher the score on the functional writing test). If the analysis of covariance is to be effective, there must be some relationship

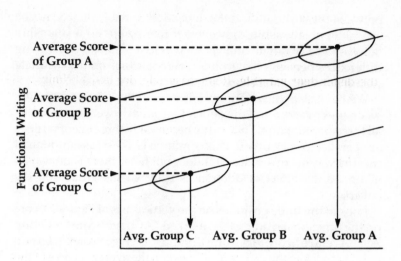

Figure 1. Reading Test Score

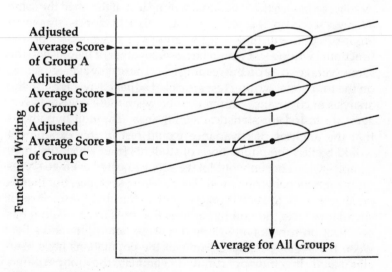

Figure 2. Reading Test Score

between the covariate and the dependent variable. It need not be a positive relationship, however. If there were no relationship between the covariate and the dependent variable, eliminating differences between the groups' average scores on the covariate would not affect their scores on the dependent variable.

We have also drawn Figure 1 as though the relationships between the covariate and the dependent variable were the same for all three populations. This is true because the three diagonal lines are parallel to each other. This condition is a basic assumption of the analysis of covariance. If it does not hold, there is no way to eliminate the effect of differences between the groups on the covariate.

Look at the three positions on the vertical axis of Figure 1 (score on the functional reading test) marked "Average Score for Group A," "Average Score for Group B," and "Average Score for Group C." Then notice the differences between the average scores of the three groups on the horizontal axis ("Avg. Group A," "Avg. Group B," and "Avg. Group C"). Part of the reason the three groups differ in their average scores on the functional writing test is because they had different average scores on the reading test, prior to instruction. Imagine what would happen to the groups' average scores on the functional writing test, if they had the same average score on the reading test. This situation is shown in Figure 2. You will find that the groups' average scores on the functional writing test would be closer together, if they all had the same average score on the reading test. These new average scores on the functional writing test are called **adjusted averages.** In the analysis of covariance, differences between these adjusted averages are tested for statistical significance. The null hypothesis that the analysis of covariance would test in this experiment would be that the populations of students from which the three samples were chosen would have equal adjusted average scores on the functional writing test. That's a long sentence, but there's no shorter way to state it precisely. It means that the analysis of covariance tests the null hypothesis that type of curriculum has no effect on functional writing test score, after differences between the average reading scores of the populations have been eliminated. It is the opportunity to eliminate the groups' differences in average pre-instructional reading score that makes the analysis of covariance more powerful than the analysis of variance.

Research and evaluation studies that make use of the analysis of covariance do not always include tests of the hypothesis that the covariate is related to the dependent variable. Nor do they always include tests of the hypothesis that the relationship between the covariate and the dependent variable is the same in all populations being compared. Regardless of their tenability, these hypotheses are often assumed to be true, and the only hypothesis tested concerns the equivalence of adjusted population average. This practice is unfortunate, since the appropriateness of analysis of covariance, and the accuracy of its result, depends critically on the correctness of its assumptions.

In the next section we consider an actual evaluation study that used the analysis of covariance, together with other analytic procedures, to examine the effect of two methods of teaching reading on post-treatment reading achievement scores.

A Comparison of Methods for Evaluating the Effect of Student Grouping Procedures

Purpose of the Study. The measurement of change in achievement in response to some instructional treatment is a problem that has plagued evaluators and researchers for decades. An entire book (Harris, 1963) convinced many researchers that there was no good way of addressing the problem unless subjects had been randomly assigned to experimental and control groups.

In this study, Williams et al. (1972) compared the results of three statistical methods in trying to determine the relative effectiveness of two procedures for grouping elementary-school students for reading instruction. They compared the analysis of gain scores (the difference between achievement at the end of the school year and achievement at the beginning of the school year); the analysis of covariance, with beginning-of-year achievement scores used as a covariate; and the analysis of so-called "residual gain scores." In the latter method, regression analysis is used to predict end-of-year achievement from beginning-of-year achievement. The residual scores from the regression analysis (that part of post-treatment achievement that is not predictable from pre-treatment achievement) are then used as the dependent variable in an analysis of variance.

The authors' primary purpose in this study was to determine whether the three methods of analysis would lead to different conclusions on the relative effectiveness of two procedures for grouping students for reading instruction.

Evaluation Design. To complete their comparison of statistical procedures, Williams et al. (1972) collected achievement test data on 165 second- through sixth-grade students in eight rural North Dakota elementary schools. The teachers used in the study were interns from the New School of the University of North Dakota, and students as well as interns were assigned to one of two reading instruction treatments. The experimental procedure consisted of instruction in "vertically grouped" reading classes, and the control procedure consisted of reading instruction in traditional, graded classes. In their data tables, the authors refer to those who received the control treatment as the "Graded Group." Vertical grouping is not well defined in the Williams et al. report, but presumably consisted of homogeneous grouping of students on some basis other than age or grade level. The method used to assign students to treatments is not described.

Students participating in the study completed achievement tests and attitude inventories both prior to instruction and at the end of the school year. Pretests, consisting of the vocabulary and reading comprehension subtests of the California Reading Test and two subtests of the Attitudes Toward Reading Inventory, were administered in October, 1970. The same instruments (or, perhaps, parallel forms of these instruments) were administered in May, 1971. Thus a period of eight months separated the pretest and posttest administrations. The authors provide no details on the forms of the tests used, nor on the procedures used in test administration. Since the instruments were standardized, one might presume that the publishers' administration procedures were followed.

Williams et al. completed separate analyses of data resulting from the administration of each subtest to students at each grade level. Since their analyses and methods of interpretation are parallel, we will consider only their results for the Reading Comprehension Subtest administered to fifth-grade students.

Authors' Data Table. Table 9 of the Williams et al. report contains a comparison of an analysis of variance on raw gain scores, an analysis of covariance, and an analysis of variance on residual gain scores on the Reading Comprehension Subtest of the California Reading Test, for 27 fifth-graders. The table is reproduced below.

Summary Data Relating to
Fifth Grade Comprehension Scores
(Authors' Table 9)

| | Comprehension Scores—Grade 5 (N=27) | | | | |
	Pre-test	Post-test	Adjusted Mean	Raw Gain	Residual Gain
Vertical Group	6.330	7.070	6.626	.740	.149
Graded Group	5.394	6.053	6.314	.659	−.087
F	8.124**	10.778**	2.031	.159	1.501
R	.495	.549	Full .864	.079	.243
R-			Rest .851		
Squared	.245	.301	Full .747	.006	.059
SS-					
Total	22.485	21.617	5.954	6.567	5.952

Critical value for significance at .05 level with df=1,25 is 4.22. **Significant at .01 level. Critical value for significance at .01 level with df=1,25 is 7.77.

Authors' Interpretation. Williams et al. do not provide detailed interpretations of the results presented in each of their 16 data tables. They chose instead to interpret the overall patterns of relationships found in the entire set of tables. They note that the three statistical methods they used to examine the effects of grouping on reading achievement often provided somewhat different results, and occasionally provided conflicting results. For example, in one instance the vertically graded group was favored by the analysis of raw gain scores, while the traditionally graded group was favored by the other two methods of analysis. In general, they found that the three methods of analysis provided results that differed in magnitude but were identical in direction. That is, when the vertically graded group or the traditionally graded group was favored by one method, it was generally favored by all three methods.

The Nature of the Inferences. Table 9 in the Williams et al. report contains the results of several inferential statistical analyses. Only one set of results pertains to the analysis of covariance, but the others are equally worthy of discussion. The rows of the table labeled "Vertical Group" and "Graded Group" contain sample statistics for 27 fifth-graders. We will discuss those data first. In the column headed "Pre-test" we find the average pre-instruction reading comprehension scores for fifth-graders who were instructed in the Vertical Group (6.330) and for fifth-graders who were instructed in the traditional Graded Group (5.394). Note that the Vertical Group began the experiment with an advantage of almost an entire point. Presumably the score scale used is grade equivalent units, but the authors do not discuss this point. In the "Post-test" column we find that the Vertical Group, with an average post-instruction score of 7.070, maintained its one-point advantage over the Graded Group (post-test average of 6.053). The column headed "Adjusted Mean" is most important to the analysis of covariance. The scores of 6.626 for the Vertical Group and 6.314 for the Graded Group are estimates of what the average post-test scores would have been for these groups, had they begun the experiment with equal average pre-test scores. These estimates are based on the supposition that all of the analyses of covariance assumptions are valid for the data in this experiment. They suppose, for example, that the relationship between pretest score and posttest score is the same, in the populations from which the Vertical Group and the Graded Group were sampled. They also suppose that the variance of posttest scores is identical in both populations. If these suppositions are not valid, the results of the analysis of covariance can be very misleading.

The column headed "Raw Gain" contains the sample average differences between posttest scores and pretest scores for the two groups. The number .740, reported for the Vertical Group, is merely the difference between the average posttest score of that group (7.070) and the average pretest score of that group (6.330). The value .659 shows that the sample of students in the Graded Group gained somewhat less than seven-tenths of a grade equivalent unit in reading comprehension, between the pretest administered in October, and the posttest administered the following May. In terms of raw gain scores, the Vertical Group appears to have outscored the Graded Group by about .08 grade equivalent units, a decidedly trivial amount.

The "Residual Gain" column contains differences between the actual average posttest score for each group, and the average of the posttest scores that were predicted from the group's pretest scores. The value .149 for the Vertical Group shows that its posttest average is slightly higher than was predicted from its students' performances on the pretest. The $-.087$ for the Graded Group shows that its students' average performance on the posttest was slightly lower than was predicted from their pretest scores. In both cases, the residual gains are so close to zero that the between-group difference would have to be regarded as trivial.

The row that is labeled "F" contains the basic inferential statistical information in the table. The value 8.124 in the Pre-test column is marked "**" to show that an analysis of variance on the pretest averages of the two groups resulted in a finding that the difference between the groups' pretest averages was statistically significant at the .01 level. In short, the Vertical Group began the experiment with a significantly higher average pretest score. The F of 10.778 in the Post-test column shows that the two groups also had a statistically significant difference between their average posttest scores. The "**" attached to the number 10.778 shows that an analysis of variance on the average posttest scores led to rejection of the null hypothesis that the two samples had been chosen from populations with identical averages. The reported F-value is significant at an alpha level of 0.01.

The value 2.031 in the Adjusted Mean column is the F-statistic that resulted from the analysis of covariance. It is the result of a test of the null hypothesis that the two groups had significantly different average posttest scores, after adjusting for the difference between their average pretest scores. In this case, the null hypothesis was retained, despite the earlier finding that the two groups differed significantly in both average pretest and average posttest scores. It is important to notice that the two groups have adjusted mean posttest scores that are much closer together (they differ by only .3 grade equivalent units) than are their unadjusted average posttest scores (which differ by more than a grade equivalent unit). By assuming that the two groups had identical pretest scores, the adjustments in average posttest score resulting from the analysis of covariance have provided a picture of what might have happened, had the groups begun the study on an equal footing. As we have noted, the validity of the adjustment depends heavily on the appropriateness of the assumptions of analysis of covariance in this study. Use of the analysis of covariance is not recommended when subjects have not been assigned to groups on a random basis, since its assumptions are often violated in such situations.

Because the Vertical and Graded groups differed so much in average pretest scores, it is likely that students were not assigned to these groups randomly. The analyses we have discussed so far point out the importance of looking beyond mere differences in average posttest scores when groups have not been assembled through random assignment. It appears that the significant posttest difference found for these groups of fifth-graders was due solely to their initial differences at the time the study began.

The last observation is reinforced by the remaining values in the row labeled "F." The .159 in the Raw Gain column is the result of an analysis of variance on the differences between the average posttest and pretest scores for the two groups. The small value of F is not statistically significant, and shows that the significant difference between the groups' average posttest scores is explained totally by their pretest differences. Likewise, the 1.501 value in the Residual Gain column is the result of an analysis of variance on the residual gain scores for the two groups. The value is not statistically significant at the 0.05 level, again showing that the significant posttest difference was due to the groups' average pretest difference.

The remaining values in Table 9, in the R, R-squared, and SS-total rows, are not pertinent to an interpretation of the analyses of variance, or the analysis of covariance. We will therefore ignore them. Don't despair, since the R and R-squared statistics are discussed in the next section of this chapter, and sums of squares have already been discussed in the chapters on one-way and two-way analysis of variance (Chapters 12 and 13).

Our Interpretation of Findings. We agree with Williams et al. that the three statistical methods used in their study produced generally similar results. However, the significance and generalizability of this finding is limited, and we would not recomend that it be extended beyond their study. Our hesitancy is based on the authors' limited reporting on the details of their study. With little or no information on how students were selected for their study, how students were assigned to the Vertical Group and the Graded Group, how the tests and attitude scales were administered, and a host of other procedural issues, we don't know whether the Williams et al. findings should be attributed to design flaws or to the experimental treatments.

For our purposes, the Williams et al. (1972) study provides a useful illustration of the analysis of covariance. We advise you to limit your consideration of their findings accordingly.

Regression Analysis

The most elementary form of regression analysis is called **simple linear regression.** One objective of simple linear regression is to predict a person's score on a dependent variable from knowledge of their score on an independent variable. Simple linear regression is also used to examine the degree of linear relationship between an independent variable and a dependent variable. In that application it is an extension of correlation analysis, which was described in Chapter 4.

Let's consider a hypothetical application of simple linear regression for purposes of prediction. Suppose you were the personnel director of a company that built high-technology devices, such as personal computers. One of your tasks was to establish policies for hiring people to work on the company's production line. Management wanted workers hired only if they could produce at least 40 modules per hour, thereby keeping up with the standard speed of the production line. Your job was to determine which prospective employees were likely to be sufficiently productive, and which were not, so that the company could avoid training workers who would later have to be fired.

When applicants for production line jobs sought employment they were given a standardized Test of Assembly Speed (TAS), which had been shown to correlate highly with employee productivity on the line. Your task was to determine the minimum score on the test that would result in an employee who was fast enough on the job. In problems like this, the minimum score on a test that is associated with satisfactory performance on the job is often called a **cutoff score.**

To determine the cutoff score, you might do the following. First, you would administer the test (TAS) to a fairly large group of applicants for production line jobs (about 200 would probably be sufficient), but you wouldn't use their scores as a determiner of whether or not they were hired. In other words, you would hire as many of the 200 as you could, basing your hiring decisions on factors other than their test scores. Next you would monitor the productivity of these employees as they worked on the production line. You would carefully record the number of modules produced per hour by each employee, until you had good estimates of the productivity of each one. You would use these production data, together with the TAS scores, to determine how well the TAS predicted worker productivity. In this problem, TAS score would be an independent variable, and the productivity measure would be a dependent variable.

Instead of just computing the correlation coefficient between the TAS scores and worker productivity, you would plot a scatter diagram of the TAS scores and productivity figures, just as was done in Chapter 4. You would then use simple linear regression analysis to determine the straight line that did the best job of going through the scatter of points. The line that fits a distribution of points in a scatter diagram better than any other line is called a **regression line**. It is determined mathematically, not by visual estimation. Many computer programs can be used to fit regression lines to the points in a scatter diagram, and some of the more sophisticated hand calculators also calculate regression lines when they are given the values of the points in a scatter diagram.

Once the regression line had been determined from the points in the scatter diagram, it would be a simple matter to find the score on the TAS that was associated with the minimum acceptable production rate of 40 modules per hour. One solution to the personnel manager's problem would then be to hire future applicants who earned scores on the TAS that were at least as high as the one that predicted a 40 module per hour production rate, and to refuse future applicants with lower TAS scores. There are other considerations that complicate the hiring problem, but we'll ignore them for the present.

Figure 3 illustrates the problem we've just described. Note that TAS score is plotted along the horizontal axis, and productivity rate (modules produced per hour) is plotted on the vertical axis. The regression line slants upward to the right. This means that, on average, applicants with higher TAS scores turn out to be more productive workers. From the hypothetical data shown in Figure 3, you can see that a productivity rate of 40 units per hour has an associated TAS score of 24. So an applicant who earned a score of 24 on the TAS would be predicted to have a productivity rate of 40 modules per hour. Note than an applicant with a TAS score of 36 would be predicted to have a productivity rate of 50 modules per hour, and that an applicant who earned a TAS score of 12 would be predicted to have a productivity rate of only 30

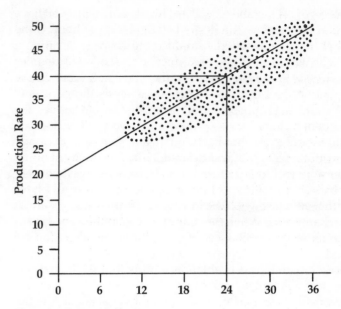

Figure 3. Score on the TAS

modules per hour.

We mentioned earlier that there were other considerations that complicated the problem of setting a cutting score on the TAS. If you look at Figure 3 carefully, you may notice that there are some workers with productivity rates above 40 modules per hour who earned TAS scores that were less than 24. There are also workers with TAS scores above 24 who have productivity rates below 40 modules per hour. This happened because the correlation between TAS score and productivity was less than 1.0. In other words, TAS score does not do a perfect job of predicting worker productivity. In this regard, our hypothetical example is typical of relationships that exist in real life. You will rarely, if ever, find an independent variable that correlates perfectly with a dependent variable. In the problem at hand, the less-than-perfect relationship between TAS score and worker productivity poses a seeming dilemma. If the personnel director recommends that production workers be hired in the future only if their TAS scores are at least 24, some applicants who would be sufficiently productive on the job will not be hired. Other applicants who have

TAS scores of 24 or above will be hired, and will later have production rates below the desired 40 modules per hour. If the personnel manager wanted to reduce the chances of hiring a worker who would later have a production rate below 40 modules per hour, (s)he should recommend that the cutoff score on the TAS be set above 24. Using regression analysis the personnel director could determine a cutoff score that would reduce the probability of hiring a worker who would have a production rate below 40 modules per hour to ten percent, five percent or any other low probability that was desired. If the probability of hiring a worker who would later turn out to be too slow was reduced, the probability of failing to hire an applicant who would have been sufficiently fast would be increased. The two kinds of errors are interchangeable. When the chances of committing one kind of error are reduced, the chances of committing the other kind are increased.

In most research and evaluation reports that use regression analysis, you will find an equation that expresses the mathematical relationship between the independent variable and the dependent variable. In the case of simple linear regression, this **regression equation** will have two parts on its right-hand side. The first will be a number that will stand by itself, and the second will be a number that multiplies the value of the independent variable. The first number will tell you what the predicted value of the dependent variable would be, if a person had a score of zero on the independent variable. The number that multiplies the value of the independent variable tells you how much the dependent variable would be predicted to change, if the value of the independent variable were to be increased by one unit. In the example we have been discussing, the regression equation would be: Predicted Productivity = 20 + 0.833 (TAS Score). The number 20 shows that an applicant who scored zero on the TAS would be predicted to have a production rate of 20 modules per hour. The number 0.833 shows that for every increase of one point on the TAS, the applicant's predicted production rate would increase by 0.833 modules per hour. How these numbers were determined is not important here, but the interpretation of the numbers is central to this discussion.

Using a regression equation, a researcher or evaluator can find the predicted value of a dependent variable for any value of an independent variable. For example, the predicted productivity rate of an applicant who scored 15 on the TAS would be 20 + (0.833)(15), which equals 20 + 12.5, or 32.5.

Other statistics that can be computed in a regression analysis tell a researcher or evaluator how well an independent variable predicts the value of a dependent variable, and how accurately an individual's score on the dependent variable can be predicted. One statistic that indicates goodness of prediction is called the **coefficient of determination.** In simple linear regression analysis, it is equal to the square of the correlation coefficient between the independent variable and the dependent variable. If in our example TAS score and worker productivity had a correlation of 0.8, the coefficient of determination would be (0.8) squared, which equals 0.64. We would then say that 64 percent of the variation in workers' production rates could be predicted by knowing their TAS scores. If TAS scores and workers' production rates had a correlation of 1.0, the coefficient of determination would also equal 1.0. We would then say that TAS score predicted 100 percent of the variation in workers' production rates. This would mean that knowing an applicant's score on the TAS would allow us to make a perfect prediction of that person's production rate as a worker. In practice, perfect prediction is never found. However, knowing a person's score on an independent variable often allows a researcher to do a far better job of predicting the person's score on a dependent variable than would otherwise be possible.

Conceptually, **multiple regression analysis** is a simple extension of simple linear regression. In multiple regression analysis, a person's score on a dependent variable is predicted from their scores on more than one independent variable. In fact, any number of independent variables can be used to predict scores on a dependent variable.

We could expand our simple linear regression example to a multiple regression problem merely by administering two tests to applicants, instead of one. In addition to the TAS, we could administer a test of persistence in the face of repetitive work. Let's call the new independent variable the Test of Worker Persistence (TWP). If the TAS and the TWP were both used in selecting applicants for production line work, the scenario would be very similar to the one we've already described. A sample of applicants would be given both tests, but their scores would not be used in determining whether or not they were hired. Their productivity on the job would be measured, just as in the simple linear regression case. Their scores on the two tests, together with their production rates, would be used to compute a regression equation that expressed the relationship between the dependent variable (production rate) and the two independent variables (TAS score and TWP score). The computations necessary to determine the regression equation would be complicated, but they need not be of concern here. A computer could be used to calculate the regression equation in very little time.

Once the regression equation had been determined, the personnel director could set cutoff scores on both the TAS and the TWP so that future applicants whose predicted production rates were below 40 modules per hour would not be hired. In fact, different combinations of TAS and TWP scores could be used in screening applicants because someone with a low score on one test could still have a predicted production rate above 40 modules per hour, provided his/her score on the other test was high enough. This is illustrated by the following hypothetical multiple regression equation.

Predicted Productivity = 15 + 0.8(TAS) + 0.6(TWP)

In this equation, an applicant who scored zero on both the TAS and the TWP would have a predicted production rate of 15 modules per hour. For every increase of one point on the TAS, an applicant would have a predicted increase in production rate of 0.8 modules per hour. And for every increase of one point on the TWP, an applicant would have a predicted increase in production rate of 0.6 modules per hour. An applicant who scored 20 on the TAS and 15 on the TWP would have a predicted production rate of 40 modules per hour: 15 + (0.8)(20) + (0.6)(15) equals 15 + 16 + 9, which equals 15 + 25, or 40. A predicted production rate of 40 would also be found for an applicant who scored 12.5 on the TAS and 25 on the TWP, since 15 + (0.8)(12.5) + (0.6)(25) is equal to 15 + 10 + 15, which equals 40.

As was true for simple linear regression, it is possible to compute statistics in a multiple regression analysis that indicate how well scores on a dependent variable can be predicted from scores on the independent variables. Two statistics that often accompany the report of a multiple regression analysis are the **coefficient of multiple correlation** (usually denoted by a capital R), and the **coefficient of multiple determination** (usually denoted by R-squared—R^2). The multiple correlation coefficient indicates how well the predicted values of a dependent variable correlate with the actual values of the dependent variable. The higher that correlation, the better the prediction. The coefficient of multiple determination is interpreted just like the coefficient of determination in simple linear regression. If the coefficient of multiple determination in our hypothetical example were equal to 0.81, it would mean that 81 percent of the variation in worker's productivity rates could be predicted by knowing their scores on the TAS *and* their scores on the TWP. As was the case in simple linear regression, perfect prediction (a coefficient of multiple determination equal to 1.0) is desired but never achieved in practice. In most employment screening applications, a coefficient of multiple determination that was larger than 0.5 would be considered quite good.

We will now consider a research study in which multiple regression analysis was used to determine how well a dependent variable could be predicted from a number of independent variables, and which of the independent variables were the best predictors. In many ways, this study is typical of research that makes use of multiple regression analysis. Its objectives include both prediction of a dependent variable and analysis of the degree of relationship between the dependent variable and a number of potential predictors that should be related to the dependent variable on theoretical grounds.

Prediction of Typing Success in High School

Purpose of the Study. Pante (1980) attempted to identify one or more variables that would predict high school students' typing speed and accuracy at the end of a one-semester typewriting course. As potential predictors, she investigated measures of students' intelligence, motivation, rhythmic ability, musical background, and a specialized test of manual speed and dexterity as potential predictors. The author's objectives were to determine which predictors, individually and collectively, would do the best job of predicting end-of-semester typewriting speed and accuracy, and to determine the strength of the relationship between each potential predictor and both the dependent variables.

Research Design. Students in six beginning typewriting classes were given three predictor tests at the beginning of their first semester of instruction. The 109 students in these six classes completed the Seashore Rhythm Test, the Digit Recall Test (used here as a measure of intelligence), and the Tapping Test (a measure of manual speed and dexterity that requires subjects to rapidly make a dot inside each of a pre-printed series of small circles). The tests were administered in the students' regular classrooms, during a single 55-minute class period, under the supervision of a trained test administrator.

In addition, students participating in the study were asked to complete a questionnaire that indicated whether or not they played a musical instrument, and whether or not they had completed a previous typewriting course of at least six weeks duration. About 50 percent of the students responded affirmatively to the latter question.

Teachers' records were used to determine the number of times each student had been absent from the typing class and the number of assignments (out of a total of eighteen) each student had completed. These variables were used as indicators of student motivation.

The dependent variables used by Pante (1980) were typing accuracy and speed scores obtained on three successive days during the last week of the semester. For purposes of analysis, each student's typing speed score was computed as the average of the speed scores earned in three three-minute tests administered on successive days. Similarly, an accuracy score for each student was determined by averaging the accuracy scores earned on the three timed tests.

To determine which of her seven independent variables, singly and in combination, did the best job of predicting typing speed and accuracy, Pante conducted a series of multiple regression analyses. She completed separate analyses using the speed and accuracy scores as dependent variables: using the data for all sampled students, using the data only for students who had less than six weeks of previous typing instruction, and using the data for students who had at least six weeks of previous typing instruction. Since the method of analysis was identical for all three groups and both dependent variables, we will consider only the prediction of typing speed, using data for all sampled students.

Author's Data Table. Table 1 in the Pante (1980) study contains the results of a multiple regression analysis in which seven independent variables were used to predict the end-of-semester typewriting speeds of 109 high school students. The table is reproduced below, just as it appeared in her report.

Pante's Table 1 actually contains the results of several regression analyses. She began with a simple linear regression model, in which typing speed was predicted only from number of student assignments completed. She then used a two-variable multiple regression model, in which the independent variables were number of assignments completed and previous typing experience. Then a third model was used, in which students' scores on the Seashore Rhythm Test were added as a third predictor. Additional models added the independent variables shown in Pante's Table 1, one at a time, in the order listed in the table. The order in which the variables were added to the model was determined on statistical grounds. Number of assignments completed was found to be the predictor variable that had the highest correlation with typing speed. That is why it was used as the independent variable in a simple linear regression model. Given that number of assignments completed was used as a predictor, previous typing experience was found to be the variable that added most to the prediction of typing speed. That is why it was added to the model next, as the second predictor in a two-variable multiple regression equation. The other variables were added to the prediction equation in the indicated order because they contributed most to the prediction of typing speed, given that the variables listed above them had already been added to the model. This type of multiple regression analysis is called **stepwise multiple regression.** Most regression computer programs allow a researcher to complete a stepwise analysis automatically. The computer determines the order in which the variables will enter the prediction equation, by first using the independent variable that has the highest correlation with the dependent variable, and then adding predictors one at a time, according to their contribution to the multiple correlation coefficient, given the earlier decisions on independent variables. This process allows a researcher to determine how much each independent variable adds to the prediction of the dependent variable, once the preceding variables have been used.

Author's Interpretation. Pante concludes that "the most important variable in predicting speed was assignments completed" (p. 23). She goes on to note that, although "all of the

Typewriting Speed of Total Students Group
(n=109)
(Author's Table 1)

Order of Variable Entry	Simple R	R-Sq.	R-Sq.Chg.	F Ratio	Degrees Freedom
1. Assignments Completed	.43*	.18	.1825	23.88*	1,107
2. Previous Typing Exper.	.42*	.31	.1320	24.31*	1,106
3. Seashore Rhythm Test	.06	.32	.0093	16.76*	1,105
4. Tapping Test	.20**	.33	.0036	12.66*	1,104
5. Digit Recall Test	−.08	.33	.0026	10.14*	1,103
6. Absences	−.16	.33	.0003	8.38*	1,102
7. Musical Instrument	.03	.33	.0002	7.12*	1,101

other predictor variables were significant at the .01 level," only previous typing experience added significantly to the prediction of typing speed, once number of the assignments completed had been used as a predictor.

The Nature of the Inferences. As we have noted above, Pante's Table 1 contains results for seven regression analyses. Because she used the stepwise regression procedure, independent variables were added to the regression equation one at a time. Each row of her table therefore contains results for a different regression model.

The column headed "Simple r" is a list of the correlations between each of the independent variables, and the dependent variable (typing speed). Notice that Assignments Completed has the highest correlation (.43), and that Previous Typing Experience is next highest (.42). These two variables, and students' scores on the Typing Test, had statistically significant correlations with typing speed. Although the asterisks in Table 1 are not explained in a footnote, the magnitudes of the correlation coefficients suggest that one asterisk denotes statistical significance at an alpha level of 0.01, while two asterisks denote significance at the 0.05 level. This notation is the opposite of tradition, but no other interpretation makes sense.

The "R-Sq." column contains values of the coefficient of determination associated with each regression analysis. As we noted above, the coefficient of determination can be interpreted as the proportion of variance of the dependent variable that is predicted by the independent variables in the regression equation. From Pante's Table 1, we can see that Assignments Completed alone predicted .18 (18 percent) of the variance of students' typing speed. When Previous Typing Experience was added to the regression equation, the two predictor variables accounted for 31 percent of the variance of typing speed. The increase in predictability gained by adding Previous Typing Experience to the regression equation is given in the "R-Sq. Chg." column. It is the difference between .31 and .18. Carried to four digits, the explained increase in variance is .1320 (a bit more than 13 percent).

Notice that 31 percent of the variance of typing speed is predicted by the first two independent variables, and that only 33 percent of the variance of typing speed is predicted by all seven independent variables. The .33 value in the R-Sq. column in the last row of the table indicates this last fact. The regression analysis therefore shows that only the first two independent variables were useful in predicting typing speed. This conclusion is reinforced by the small values in the R-Sq. Chg. column, for all independent variables beyond Previous Typing Experience. The R-Sq. Chg. value of .0093 for the Seashore Rhythm Test indicates that it adds less than one percent to the predictable variance of typing speed, once the first two independent variables have been used as predictors. The remaining independent variables add even less, as indicated by their small values in the R-Sq. Chg. column.

The numbers in the "F Ratio" column are sample F-statistics that would be used to test the null hypothesis that the independent variables in the regression equation at each step in the analysis contributed *nothing* to the prediction of typing speed. For each of the seven regression models, this null hypothesis was rejected at the 0.01 level of significance (as denoted by the single "*" attached to each F value). The numbers in the "Degrees Freedom" column indicate the appropriate degrees of freedom for determining the significance of the reported F-statistics.

It is important to realize that Pante did not tabulate any of the actual regression equations that resulted from her study. This is because she was interested in determining how well students' typing speed could be predicted using a variety of independent variables, and was not interested in predicting the typing speed of any individual student. One might describe her research objectives as "explanation" rather than "prediction," since the principal objective of her study was determining how much of the variance of the typing skill variables could be "explained" by her seven independent variables.

Our Interpretations and Conclusions. We agree with Pante's interpretation of the results in Table 1. Her conclusion that it is possible to predict students' typing speed from an indicator of their motivation (number of assignments completed) is sound. However, you should note that this variable only accounted for 18 percent of the variance of typing speed, and none of the other independent variables used in the study (other than previous typing experience) contributed in a practical way to the prediction of typing speed. Thus the greater part of the variance of typing speed was not predicted by any of the independent variables Pante (1980) examined.

Pante's study was unusually well documented. It provides a good illustration of the use of multiple regression analysis when explanation of the relationships among a dependent variable and a set of independent variables is of interest. The kinds of statistics reported in this study are typical of those you will find in many reports on multiple regression analyses. When the researcher's objectives include prediction as well as explanation, you will also find the kinds of regression equations we described above.

Factor Analysis

Factor analysis is used extensively in research and evaluation studies that involve the development of new measurement instruments. It is particularly useful as a tool for examining the validity of tests or the measurement characteristics of attitude scales.

The principal objective of factor analysis is to construct a small number of variables (called **factors**) that do a good job of conveying the information present in a large number of variables. For example, a 30-item attitude scale might be designed to secure information on students' feelings toward their school, toward their teachers, and toward academic work in general. In the minds of its developers, the attitude scale should measure three attitudinal factors (attitudes toward school, teachers, and academic work) even though it is made up of 30 variables (the individual items in the scale). Using factor analysis, it would be possible to determine whether the 30 items really form three distinct attitude factors, or whether the items measure a larger or smaller number of attitudes. It would also be possible to identify the specific items associated with each factor.

Computationally, factor analysis is a very complex statistical procedure. Most factor analyses would be impractical without the use of a computer. Fortunately, a number of widely-available computer programs allow applied researchers and evaluators to complete factor analyses easily and with impressive speed. Although we need not concern ourselves with the computational details of factor analysis, some time spent considering its underlying logic will aid in understanding its purposes and results.

In our example involving a 30-item attitude scale, students' responses to each of the 30 items would define a separate variable. Each student who completed the attitude scale would have 30 scores, and each of these scores would vary across a sample of students. A factor analysis of the attitude scale would begin by calculating the correlation coefficient between every pair of items. From our discussion of correlation in Chapter 4, you might remember that two variables have a correlation of zero if scores on one have no relationship to scores on the other. If two items on the attitude scale had a correlation of zero, it would mean that students' responses to one of the items would not be linearly related to their responses to the other item. It would be illogical to regard such items as two indicators of the same kind of student attitude; the items would have to represent different attitudes. Conversely, students' responses to another pair of items on the attitude scale might have had a correlation of 1.0. The new pair of items would apparently be measuring the same kind of attitude, since higher student responses to one of the items would always correspond to higher responses to the other item. A single variable would convey as much information about students' responses to this pair of items, as would the two original variables. That single variable could be called a "factor." Its meaning would be determined by the content of the pair of items it represented.

In practice, pairs of items on an attitude scale will rarely have correlations of zero or 1.0. Most correlations will be somewhere in between. The higher the correlation between students' responses to a pair of items, the more reasonable it would be to think of those items as indicators of a single underlying factor. If the correlation between a pair of items were less than 1.0, no single factor could perfectly represent the items. But if the correlation were high enough, most of the salient information in students' responses could be expressed by their scores on a factor that was constructed from their item scores.

Factor analysis considers the correlations between *every pair* of items on a scale or test. For example, if a scale consisted of only four items numbered 1, 2, 3, and 4, factor analysis would begin with the correlations between Items 1 and 2, Items 1 and 3, Items 1 and 4, Items 2 and 3, Items 2 and 4, and Items 3 and 4. These six correlation coefficients would be analyzed to determine whether the items could be grouped in such a way that all items in a group had fairly high correlations with each other. Each group of items would then be a good candidate for representation by a single factor.

The literature on factor analysis includes a variety of procedures for deriving factors from sets of variables. If you were to read one of the many texts devoted exclusively to factor analytic methods, you would find such terms as **principal components analysis, principal factor analysis, image analysis, alpha factor analysis,** and **Rao's canonical factoring method.** Each of these factor analysis procedures has the same basic objective—to determine factors that do an adequate job of conveying the essential information in a larger set of variables. The procedures differ in their underlying assumptions and their computational details. In the educational research and evaluation literature, principal components analysis and principal factor analysis appear to be most widely used. Principal components analysis assumes that each of the original variables can be divided into two parts: an error component that reflects the less-than-perfect reliability of all educational measurements, and a **true-score** component that is common to all of the variables being factor analyzed. Principal factor analysis assumes that each of the original variables can be divided into three parts: an error component, a true-score component that is common to all of the variables being factor analyzed, and a true-score component that is unique to each variable.

These sets of assumptions will be plausible in different applied problems. Plausibility should be the principal guide in selecting a factor analysis procedure.

After the correlations between every pair of variables have been computed, both principal components analysis and principal factor analysis determine a set of factors called an **original solution** or an **unrotated solution.** In an unrotated solution, an initial factor that does the best job of representing all of the variables is determined first. Then a second factor that does the next best job of representing all of the variables is found. The second factor is required to be uncorrelated with the first. Next a third factor that is uncorrelated with the first two factors is found. This factor must also do the best job possible of representing the information in all the variables, subject to the requirement that it cannot be correlated with the first or second factors. Principal components analysis and principal factor analysis proceed in this way, until the number of factors determined is equal to the number of variables being factor analyzed.

Figure 4, below, is an attempt to illustrate the first two factors that might be found in a principal components analysis or a principal factor analysis. In this figure the short arrows represent the original variables to be factor analyzed. One group of arrows might represent items that concern students' attitudes toward education. The other group might represent students' attitudes toward their teachers. Any two arrows that are at a right angle to each other represent variables that are uncorrelated. Any two arrows that fall right on top of each other represent variables that have a correlation of 1.0. In general, the smaller the angle between two arrows, the higher the correlation between the variables they represent. The larger arrows labeled "Factor 1" and "Factor 2" represent the first factors that might be derived in a principal components analysis or a principal factor analysis of the variables. Notice that Factor 1 falls roughly mid-way between the two groups of arrows, and that Factor 2 is at a right angle to Factor 1. The right angle means that Factors 1 and 2 are uncorrelated, as required by the factor analytic methods.

Several things we have said about the objectives and procedures of factor analysis might appear to be contradictory at this point. First, we stated that the objective of factor analysis is to determine relatively few factors that do a good job of representing a far larger number of variables. But in describing the princi-

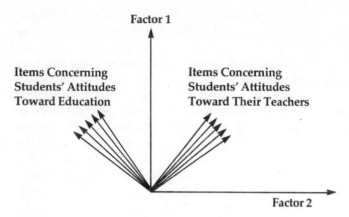

Figure 4. The First Two Unrotated Factors Resulting from a Hypothetical Principal Components Analysis of a Student Attitude Scale

pal components analysis and principal factor analysis procedures, we said that the methods would produce the same number of factors as original variables. There appears to be a conflict here. In addition, we said that the factors are intended to *represent* the original variables. Yet the factors that are shown in Figure 4 do not appear to represent either group of variables. Neither Factor 1 nor Factor 2 is particularly close to the two groups of original variables. The correlations between the variables and the factors would appear to be moderate at best. Additional steps in the factor analysis procedure are needed to resolve these seeming contradictions. The first is called **factor selection**, and the second is called **factor rotation**. We will describe these steps briefly.

Although the number of factors needed to convey *all* of the information in a set of variables is the same as the number of variables, some smaller number of factors can usually do a good job of conveying most of the information. In a principal components analysis or a principal factor analysis it is usually found that relatively common few factors (sometimes only two, three or four) describe or "account for" most of the information in the original set of variables. The remaining factors often account for very small proportions of the information in the original variables, either by themselves or in combination. One kind of statistic that results from a principal components analysis or a principal factor analysis is a listing of the **proportion of variance** that each factor accounts for. These statistics are regarded as indicators of the relative potency of each factor in conveying the information in the original set of variables. These proportions are reviewed when factors are selected. The only factors that are used in further analyses are those that account for appreciable proportions of the variation in the original variables.

Once factors have been selected for further analysis, they are **rotated** (which we'll explain in a moment), so that they do a better job of representing the original variables. Ideally, after rotation each group of variables will have high correlations with one of the rotated factors, and low correlations with all of the others. This is the sense in which we say that a factor "represents" a group of variables.

Just as there are many different factor analysis methods, researchers and evaluators can choose from among many different factor rotation procedures. Three procedures that are used quite often in educational research and evaluation are called **Varimax, Equimax,** and **Quartimax.** Although these procedures differ in their statistical details, they have identical objectives. All seek to define a set of rotated factors that have high correlations with some of the original variables, and low correlations with all others. All require that the rotated factors be uncorrelated with each other. Factors that are uncorrelated with each other are called **orthogonal factors.** Factors that are correlated with each other are called **oblique,** since if they were graphed as in the figure above, the angles between them would be smaller than right angles. One of the most widely used rotation procedures that results in correlated (oblique) factors is called the **Direct Oblimin** method. Its objectives are similar to those of the orthogonal rotation procedures (Varimax, Equimax, and Quartimax). All rotation procedures seek to determine a set of factors that clearly represent differing subgroups of the original set of variables.

Figure 5, below, illustrates the results of rotating the first two factors shown in Figure 4. Notice that rotated Factor 1 goes right through one group of small arrows (representing items that measure students' attitudes toward education in general), and that rotated Factor 2 goes right through the other group (representing items that measure students' attitudes toward their teachers). Because of this, we would call rotated Factor 1 an "attitude toward education factor" and rotated Factor 2 an "attitude to-

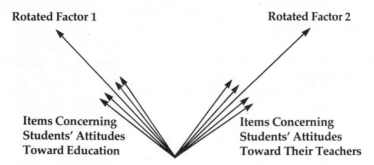

Figure 5. The First Two Rotated Factors Resulting from a Hypothetical Principal Components Analysis of a Student Attitude Scale.

ward teacher" factor. In this example, as in all factor analyses, the meaning that is attached to a factor derives from the variables that are highly associated with it.

Because we constructed Figure 4 so that two groups of arrows clearly represented distinct sets of variables, and because the groups of variables were centered around arrows that were at right angles to each other, we were able to illustrate two orthogonal factors that do a good job of representing the variables. Therefore an orthogonal rotation procedure, such as Varimax, Equimax, or Quartimax, would have been appropriate in this example. Had the two groups of variables not been at right angles to each other, an oblique rotation procedure, such as the Direct Oblimin method, would have been needed to define factors that did a good job of representing subgroups of variables.

As you read research and evaluation reports that include factor analyses, you will come across additional specialized terms. For example, authors will discuss the **loadings** of the variables on the factors. Although the definition is strictly correct only in the case of orthogonal factor analyses, you can think of factor "loadings" as correlations between the variables and the factors. It will always be the case (whether a factor analysis results in orthogonal factors or oblique factors) that the higher the correlation between a variable and a factor, the higher the loading of that variable on the factor.

Other terms that arise frequently in discussions of factor analysis results are **communality** and **eigenvalue.** Communalities are associated with the original variables that are to be factor analyzed. The communality of a variable is the proportion of its total variance that is shared with (or common to) the other original variables. The higher the correlations among the original variables, the more they have in common, and the higher their communalities. Communalities can range between zero and 1.0. A variable with a communality of 0.6 would be described as having 60 percent of its variation in common with the other variables to be factor analyzed. If a variable has a very low communality (close to zero), it is not likely to be associated with any other variables in defining a factor. The second term we highlighted earlier was "eigenvalue." Eigenvalues are associated with factors. They indicate how much of the variation in the entire set of original variables is "accounted for" by each factor. One rule that is often used to select the factors in an original solution that will be *eliminated* from further analysis is to discard those that have eigenvalues smaller than 1.0. In that way, only factors that convey a non-trivial proportion of the information in the original variables are retained for interpretation. Because of the way factors are defined in a principal components analysis or a principal factor analysis, the first factor will always have the largest eigenvalue; the second factor the next largest eigenvalue, etc. Since the eigenvalues associated with succeeding factors will be progressively smaller, a decision to ignore some of the latter factors is often justifiable.

As in earlier portions of this chapter, we will now describe an actual research study that used the statistical procedures we have discussed. In the description that follows, notice that the study is similar in several ways to the hypothetical factor analysis of the 30-item attitude scale we described above. A small number of factors that do a good job of representing the items in a measurement instrument were sought, both in the hypothetical study and in the real one that follows.

A Factor Analysis of an Instructor Rating Scale

Purpose of the Study. Jaeger and Barnes (1981) compared three different methods of factor analyzing a rating scale used by college students to evaluate their instructors. Their objective was to determine whether the three methods would result in the same number of factors, and whether the resulting factors would be defined similarly, regardless of the method used.

Research Design. The rating scale used by Jaeger and Barnes contained 20 items. Its developers thought that it would measure students' perceptions of two related dimensions of instructor quality; they therefore hypothesized two underlying factors. The first ten items in the scale were intended to measure students' perceptions of the overall quality of their instructor. The last ten items were designed to measure students' perceptions of their instructor's conscientiousness.

In one method of analysis, the authors correlated individual students' ratings of instructors on each pair of items on the scale, ignoring the fact that students were rating 13 different instructors in 19 different classes. The correlations were based on ratings provided by 493 students. These correlations were then used to complete a principal factor analysis followed by a direct oblimin rotation of factors with eigenvalues greater than 1.0. Jaeger and Barnes called this procedure an "across-students" analysis.

In their second method of analysis, called a "between-classes" analysis, Jaeger and Barnes first averaged the ratings of individual students within each of the 19 classes. They then computed correlations between the averaged ratings given the instructor in each class, for all pairs of items on the rating scale. These correlations were used to complete a principal factor analysis, followed by a direct oblimin rotation of the original solution, just as in the across-students analysis.

The final method the authors used was called a "within-classes" analysis. In this procedure, Jaeger and Barnes first computed correlations between the ratings provided only by students in a given class, for each pair of items on the rating scale. A separate table of correlations was computed for each of the 19 classes. Corresponding correlations (e.g., between Items 1 and 2, Items 1 and 3, etc.) were then averaged across the 19 classes, to form a final table of correlations. These correlations were then used in a principal factor analysis, followed by an oblique rotation of the factors in the original solution, just as in the other methods.

Authors' Data Table. The three types of analysis conducted by Jaeger and Barnes resulted in three tables of factor loadings, eigenvalues, and correlations among rotated factors. Only the table for the "across-students" analysis will be shown here, since all three tables contain the same kinds of statistics. Jaeger and Barnes (1981) Table 1 is reproduced below.

Table 1
Oblique Factor Pattern Matrix Resulting from Analysis of Students' Ratings of Faculty (across students) (Only loadings exceeding 0.45 are shown)

	Factors			
Items	1. Overall Effect-tiveness	2. Instructors' Effort	3. Students Understand Standards	4. Fair, Equitable Treatment
1. Overall Effectiveness	1.14			
2. Take This Instructor	.78			
3. Take Another This Instructor	.94			
4. Recommend Instructor	.87			
5. Among Best Teachers	1.05			
6. Effect on Learning	1.18			
7. Effectiveness in Stimulating	1.03			
8. Effectiveness, Understanding	.95			
9. Stimulate Interest	1.14			
10. Professional Growth	1.10			
11. Full, Prompt Evaluation		.54		
12. Class Hours Regular		.52		
13. Evaluation Fair, Equitable				
14. Instructor Prepared		1.22		
15. Instructor Willing to Meet		.71		
16. Instructor Ridicules				.82
17. Responsive to Student Concerns				.84
18. Makes Unfair Demands				1.10
19. I Understand Grading			.68	
20. I Understand Course Purposes	.64			
Eigenvalue	10.24	1.86	1.14	1.03
Percent of Explained Variance	51.2	9.3	5.7	5.1
Cumulative Percent of Explained Variance	51.2	60.5	66.2	71.3
Factor Intercorrelations				
Factor 2	.80			
Factor 3	.44	.44		
Factor 4	.79	.83	.34	

Authors' Interpretation. On pages 7 and 8 of their paper, Jaeger and Barnes state:

> "The four factors resulting from the "across-students" and the "between-class" analyses seemed to represent the same general constructs, although the order of the second, third, and fourth factors in these solutions differed. For example, in the analysis of the across-students correlation matrix, Factor 2 ("Instructor's Effort") was defined by items on the promptness of the instructor's evaluations (Item 11), whether classes were held during regular hours (Item 12), whether the instructor was prepared to teach class (Item 14), and whether the instructor was willing to meet with students outside of class (Item 15)."

In their discussion of the correlations among the factors and the percentages of variance accounted for by the first four factors, the authors state (pages 8 and 9):

> "Factor intercorrelations are shown in Tables 1 and 2 for the across-students and the between-class analyses respectively. Comparison reveals that the correlations resulting from the between-class analysis were consistently lower than corresponding correlations for the across-students analysis. This suggests that the between-class factors were more nearly orthogonal than those resulting from the across-students analysis."

> "At least half the variance in the 20-dimensional space of the instructor rating scale was accounted for by the first factor in both the across-students and between-class analyses . . . . The four rotated factors in the across-students analysis accounted collectively for more than 71 percent of the variance of the 20-dimensional space."

The Nature of the Inferences. In contrast to the statistical procedures we have discussed in a number of earlier chapters, no formal hypothesis tests are conducted in a factor analysis. The kind of statistical inference that takes place in factor analysis is point estimation. In the process of determining factors, a number of population parameters are estimated, including factor loadings, eigenvalues, communalities, and in the case of oblique factor rotations, the correlation between every pair of factors. One could argue that factor analysis is a descriptive statistical procedure, rather than an inferential one, since the numerical values that result from a factor analysis could be viewed as descriptive statistics that apply only to the original sample. However, researchers are rarely interested in the measurement properties of tests, attitude scales, and other instruments, *only* as they apply to a particular sample. It is difficult to imagine a research or evaluation study in which the interpretation of a factor analysis would be strictly limited to a given sample of respondents or examinees. They are usually inferring the existence of a factor or factors in a wider population.

The upper portion of Jaeger and Barnes' Table 1 contains the loadings of 20 items on four rotated factors. Because the authors used an oblique rotation procedure, the factor loadings are *not* correlation coefficients. They can instead be interpreted as the coefficients of the items, were each factor to be used as a dependent variable in a multiple regression analysis, with the items as independent variables. Note that the table title states: "(Only loadings exceeding 0.45 are shown.)." This means that each of the 20 items had a loading on each of the four factors, but that small loadings (those less than 0.45) have been omitted so as to facilitate the interpretation of results.

The labels given the four factors shown in Table 1—"Overall Effectiveness," "Instructor's Effort," "Students Understand Standards," and "Fair, Equitable Treatment"—are inventions of the authors. They considered the items that had high loadings on each factor, and attempted to find a concise label that represented the content of each set of items. The process of labeling factors is an art rather than a science. In fact, some who have scoffed at factor analysis consider it to be the only creative part of the procedure. Often, it is necessary to ignore some variables or to do an inadequate job of characterizing the content of others when trying to find a concise name for a factor. It is therefore wise to do a careful review of the content of the original variables when you interpret a research or evaluation study that contains a factor analysis. For example, Jaeger and Barnes' label for their first factor ignores the item "I Understand Course Purposes," even though it has a loading of 0.64 on that factor and does not have a high loading on any other factor. It may well be that students consider their understanding of the purposes of a course when they judge the overall effectiveness of an instructor, but one would not expect such considerations to be paramount. The label given the first factor more obviously reflects items in which students rated the overall effectiveness of their instructor, considered whether or not their instructor was among the best teachers they had experienced in college, and judged their instructor's influence on various aspects of their learning, interest, professional growth, and understanding. Since items involving these content areas had the highest loadings on Factor 1, they are best represented by the factor. The factor label is therefore supportable, even though it is not totally inclusive.

Four variables had moderate to high loadings on the second factor. The variable with the highest loading, "Instructor Prepared," reflects one aspect of an instructor's effort in teaching. High ratings on the other variables that loaded on the second factor would also be characteristic of instructors that invested time and effort in their teaching, although one could certainly argue that the content of these items is not altogether clear from the label given Factor 2.

Notice that Factor 3 has only one variable (Item 19) with a moderately high loading on it. Although this often happens in a factor analysis, it is not a desirable outcome, since a factor that is defined by a single variable does not contribute to a reduction in the number of variables necessary to characterize the original set of correlations.

The fourth factor in Jaeger and Barnes' Table 1, labeled "Fair, Equitable Treatment," is defined by three variables. These variables reflect students' perception of the instructor's willingness to meet with them, *absence* of instructor behaviors that would demean or ridicule students, and *absence* of instructor behaviors that would place unfair demands on students. It would seem that these kinds of behaviors do contribute to what many would term "fair and equitable treatment" of students. However, it could be argued that the factor label is too broad, since many other aspects of instructor fairness were not represented in the rating instrument.

The second section of Jaeger and Barnes' Table 1 contains eigenvalues, in addition to rows labeled "Percent of Explained Variance" and "Cumulative percent of Explained Variance." As we noted above, eigenvalues are proportional to the percentage of the variation in the original set of variables that is "explained" or "accounted for" by the factors. Notice that Factor 1 has an eigenvalue of 10.24, and that its "percent of explained variance" is 51.2. Since there were 20 variables in the rating instrument, the eigenvalues for all factors (not just the four that were rotated in this analysis) must sum to 20. To determine the percent of variance that Factor 1 explains, one can merely divide its eigenvalue by the sum of the eigenvalues, and then multiply by 100 to convert the proportion to a percentage.

The important information that resulted from this analysis was that the first factor explained over half of the variance in the original set of items. The second through the fourth factors explained relatively small percentages of variance, since their eigenvalues are small. The second factor explained just over nine percent, and the third and fourth factors explained just under six percent and just over five percent, respectively. The "Cumulative Percent of Explained Variance" row in Table 1 tells us that, taken together, the four rotated factors explained just over 71 percent of the variance in the original variables. This means that almost 30 percent of the variance in the 20 original items is not accounted for by the four rotated factors, and must be presumed to consist either of measurement error or true variance that is unique to individual items. Whether it is sufficient to describe the relationships among a set of variables solely in terms of factors that account for only 70 percent of their variance is a matter of subjective judgment. Reducing 20 variables to only four representative factors affords great economy of description. Certainly, we cannot expect the representation to be totally accurate.

The final section of Jaeger and Barnes' Table 1 contains correlations among the four factors. It is noteworthy that several pairs of factors have correlations that are quite large. For example, the correlation between Factors 1 and 2 is .80. One might interpret this result by suggesting that Factors 1 and 2 could be replaced by a single factor, consisting of all of the variables with high loadings on both of them. If a single factor were used, the representation of the original variables would be simplified. However, the proportion of explained variance would be smaller, so that the representation of the original variables would be less complete. With a correlation of .80, Factors 1 and 2 have an angle of about 37 degrees between them. Since this angle is far from zero, the use of two factors, rather than one, seems justified.

Conclusions. Our interpretation of the Jaeger and Barnes (1981) study illustrates several important principles that apply to virtually all factor analyses. First, it was possible to "explain" over 70 percent of the variance of the 20-item rating scale in terms of only four underlying factors. This result illustrates a basic objective of factor analysis—the quest for parsimony. The result also illustrates the less-than-perfect representation of the original variables that one inevitably discovers in a factor analysis. Here, almost 30 percent of the variance of the rating scale was not explained by the four rotated factors. Second, the first factor explained over 50 percent of the variance in the original rating scale and was defined almost totally by items representing students' perceptions of the overall quality of their instructors. This result was consistent with the intentions of the scale's developers, and confirmed their hypothesis that one underlying dimension of the rating scale would be an "overall effectiveness" construct. However, the remaining items in the scale did not cluster together to define a "conscientiousness" factor, as the developers had intended. Instead, three additional factors with eigenvalues greater than one were found. Subclusters of the supposed "conscientiousness" items defined the additional factors. Thus the results of the factor analysis were only partially consistent with the developers' expectations about the relationships among their instructor rating items. These results illustrate an important application of factor analysis. The method allows researchers to examine the validity of theories that involve relationships among constructs or the relationships among operational measures and underlying constructs. In response to the partial confirmation of their hypotheses about the two factors underlying their rating scale, the developers of the scale could either revise their instrument or their theory, depending on the importance of having a two-dimensional scale.

In all of the factor analysis procedures we have described and illustrated, the essential outcomes of the analyses are prescribed more by the data than the researcher. For example, once the researcher decides to retain factors that have eigenvalues greater than one, the relationships among the original variables dictate the number of factors that will be rotated. The data also determine the pattern of loadings of the variables on the factors. Although our interpretation of the results of the Jaeger and Barnes (1981) study was logically confirmatory (since the developers of the rating scale had specific expectations about the number of factors that would be found and which item would define which factors), the method of analysis was essentially exploratory. None of the researcher's expectations about the pattern of factor loadings influenced the analysis. Some very recent factor analysis procedures, called **confirmatory factor analysis methods,** allow a researcher to specify some or all of the outcomes of a factor analysis, in accordance with underlying theory. For example, the researcher can specify the number of factors to be used, the pattern of factor loadings, or the magnitudes of the factor loadings. One outcome of a confirmatory factor analysis is a specific statistical test of the hypothesis that the data fit the hypothesized factor model. Confirmatory factor analyses have not been widely used in educational and social science research to date, but it is likely that they will see increasing application in the near future. If you find a factor analytic study that has used the **Lisrel** computer program (Joreskog and Sorbom, 1978), you can be sure that the study is confirmatory rather than exploratory.

Summary

In this chapter we have described and illustrated three sophisticated statistical procedures that are used quite frequently in research and evaluation studies. Each of these procedures is often the subject of an entire book, rather than a portion of a single chapter. Our discussion of these procedures is therefore limited and superficial. However, our objective was to introduce you to each procedure, rather than provide complete coverage. Hopefully, when you encounter research and evaluation reports that include these methods, they will appear less mysterious and more understandable. If you can "get the flavor" of the authors' arguments and understand the essence of the studies' outcomes, you've learned what we intended.

The analysis of covariance allows a researcher or evaluator to adjust for pre-existing differences between treatment groups, on one or more continuous variables, when examining the effects of a categorical independent variable on a dependent variable. The procedure thus enhances the researcher's control over the outcomes of an experiment, and strengthens the claim that an independent variable "caused" a dependent variable. The procedure is used quite frequently in studies that involve pre-treatment and post-treatment measures since it provides a way to adjust for differences between the pre-treatment status of experimental groups before examining the effects of various treatments on a post-treatment measure.

Multiple regression analysis is very useful in studies that involve an examination of the relationships between a dependent variable and one or more independent variables. Sometimes the purpose of the examination is prediction, as in situations where candidates must be selected for a limited number of positions. In these situations, the objective of multiple regression analysis is to discover which of a set of potential independent variables is useful in predicting an individual's score on a dependent variable. In other studies, the objective of multiple regression analysis is explanation rather than prediction. In these studies, the researcher wants to know how much of the variation of a dependent variable can be "explained" by one or more independent variables. Another frequent objective of such studies is the determination of the relative potency of several independent variables in predicting a dependent variable.

Factor analysis is a statistical method for determining a relatively small number of variables that "explain" the relationships among a relatively large number of variables. The newly-determined variables are called factors. One objective of factor analysis is parsimony—to represent many variables by few variables. Other objectives of factor analysis might be confirmatory (to see whether the factors that result from the analysis are consistent with some underlying theory or hypothesis) or exploratory (to find out what the data "say," without having some a-priori theoretical expectations). In either case, factor analysis is particularly useful in examining and understanding the composition of measurement instruments used in education or the social sciences.

Conclusion

While this is certainly not an easy book to read, neither is it easy to become a *knowledgeable* spectator of any serious sport. By now, you should have acquired a very good reader's understanding of quite advanced statistics and we believe it will prove not only very useful to you but interesting in its own right.

SOME ADVANCED TOPICS / 325

Problems — Chapter 14

14.1 Harbinger and Spring (1978) conducted a study on the effect of "engaged instructional time" on the reading achievement of third-grade students. Their study was a true experiment, in which they randomly assigned 16 third-graders to each of three instructional groups that were identical in every respect, except the length of time during each school day devoted to active instruction in reading. Group A was provided 30 minutes of reading instruction per day; Group B was provided 45 minutes of reading instruction per day; and Group C was provided 60 minutes of reading instruction per day. Teachers were rotated among the three groups, with each of three teachers assigned to each group for a one-week period, prior to moving on to the next group. This design feature helped to separate the effects of length of instruction from teacher effects.

The researchers administered a verbal intelligence test to all participating students at the beginning of the nine-month experiment. After nine months of instruction, they administered the Reading Comprehension subtest of the Iowa Tests of Basic Skills to all participants, as a measure of instructional effectiveness.

The researchers used the analysis of covariance to examine the effect of length of instruction on reading achievement.

    a. Based on this description of Harbinger and Spring's research, identify the following variables:
        Independent variable
        Dependent variable
        Covariate
    b. What null hypothesis were Harbinger and Spring testing in their research study?
    c. Why do you think Harbinger and Spring used the analysis of covariance in their study, instead of an analysis of variance? What advantage could they gain from an analysis of covariance?

14.2 Coleman, et al., (1966) conducted a famous study on equality of educational opportunity in the United States. Their results on the relative effectiveness of school factors and home background factors in predicting the academic achievement of elementary and secondary students have been cited by researchers, policy analysts, and politicians for eighteen years. The principal analytic method used by Coleman and his colleagues in investigating the effects of schools and the home on achievement was stepwise multiple regression analysis. Suppose that they reported the following results, with arithmetic achievement for sixth-graders as the dependent variable:

Variables (by order of entry)	Correlation	R-squared	R-squared Change	F
Mother's Education	.42	.18	.18	17.88*
Family Income	.36	.26	.08	16.06*
Hours of Arithmetic Instruction	.16	.28	.02	13.21*
Teacher's Education	.10	.29	.01	11.67*

*significant at 0.05 level.

a. Which independent variable contributes most to the prediction of sixth-grade arithmetic achievement? Support your answer by citing data in the table.
b. What percentage of the variance of arithmetic achievement is predicted by the four independent variables?
c. What null hypotheses did the authors test when they reported F-statistics that were "significant at the 0.05 level?"

14.3 In a factor analysis, what does a researcher mean when he or she reports that "Three factors accounted for 76 percent of the variance of the ten items"?

References

Alexander, L., and Brown, R. Evaluating attitudinal change in a multicultural setting. Presented at the Annual Meeting of the American Educational Research Association, Toronto, Canada, March, 1978. ED 169 086.

Campbell, P.B. Feminine intellectual decline during adolescence. Presented at the Annual Meeting of the American Educational Research Association, Chicago, IL, April, 1974. ED 091 620.

Cleveland, W.E., and Idstein, P.M. In-level versus out-of-level testing of sixth grade special education students. Presented at the Annual Meeting of the National Council on Measurement in Education, Boston, April 8-10, 1980. ED 187 725.

Downie, N.M., and Heath, R.W. *Basic Statistical Methods*. 4th ed. New York: Harper & Row, 1977.

Glass, G.V., and Stanley, J.C. *Statistical Methods in Education and Psychology*. Englewood Cliffs, N.J.: Prentice-Hall, Inc., 1970.

Harris, C. *Problems in Measuring Change*. Madison, WI: University of Wisconsin Press, 1963.

Hays, W.L. *Statistics for the Social Sciences*. 2nd ed. New York: Holt, Rinehart and Winston, 1973.

Hays, W.L. *Statistics*. 3rd ed. New York: Holt, Rinehart and Winston, 1981.

Jaeger, R.M., and Barnes, E.A. Methodological influences on the dimensions of instructor performance. Presented at the Annual Meeting of the North Carolina Association for Research in Education, Greensboro, N.C., November, 1981.

Joreskog, K.G. and Sorbom, D. *LISREL-IV*. Chicago: National Educational Resources, 1978.

Kershner, K.M., and Blair, M.W. Career education program, 1974–1975. Final evaluation report. Research for Better Schools, Inc., September, 1975. ED 138 826.

Koffler, S.L. An analysis of ESEA Title I data in New Jersey. Occasional Papers in Education, New Jersey State Department of Education, Trenton, N.J., 1977. ED 146 217.

Kovner, A., Burg, L., and Kaufman, M. An evaluation of Project LEAP, ESEA Title I Program of Medford, MA, 1975–1976. Northwestern Univ., Boston, MA, 1976. ED 156 732.

Law, Alexander. Personal communication, 1982.

McConnell, J.W. Relations between teacher attitudes and teacher behavior in ninth-grade algebra classes. Presented at the Annual Meeting of the American Educational Research Association, Toronto, Canada, March 27–31, 1978. ED 152 751.

Mendenhall, D.R. Relationship of organizational structure and leadership behavior to staff satisfaction in IGE schools. Technical Report No. 412. Wisconsin Univ., Research and Development Center for Cognitive Learning, March, 1977. ED 145 562.

Messick, S. Test validity and the ethics of assessment. *American Psychologist*, 1980, *35*, 1012-1027.

Messick, S. Evidence and ethics in the evaluation of tests. Presented at the Annual Meeting of the American Educational Research Association, Los Angeles, April, 1981.

Pante, C.M. The relationship between a linear combination of intelligence, musical background, rhythm ability and tapping ability to typewriting speed and accuracy. Presented at the Annual Meeting of the American Educational Research Association, Los Angeles, April 13–17, 1981. ED 201 780.

Philadephia School District, Get Set Day Care Program. Philadelphia, PA, 1972. ED 088 561.

Pool, K.W., and Capie, W.R. The Miltz (1971) How to Explain Program and its effects upon preservice elementary teacher explaining ability. Presented at the Annual Meeting of the American Educational Research Association, NY, April 4–8, 1977. ED 139 755.

Schwer, W.E. An evaluation of the effects of an open space program on selected seventh grade pupils and their teachers. Practicum Report, Nova University, National Ed.D. Program for Educational leaders, June, 1973. ED 088 182.

Sewell, A., Dornseif, A., and Matteson, S. Controlled multivariate evaluation of open education: applicaton of a critical model. Presented at the Annual Meeting of the American Educational Research Association, Washington, D.C., March 30–April 3, 1975. ED 109 250.

Spear, M. *Charting Statistics.* New York: McGraw-Hill, 1952.

Williams, J.D., et al. A comparison of raw gain scores, residual gain scores, and the analysis of covariance with two modes of teaching reading. Presented at the Annual Meeting of the American Educational Research Association, Chicago, IL, April, 1972. ED 061 287.

Glossary

ANALYSIS OF COVARIANCE (ANACOVA): An **inferential** statistical procedure used to test the **null hypothesis** that the **means** of three or more **populations** are equal to each other, after random samples selected from the **populations** are adjusted to eliminate differences between their average values on one or more independent variables (called covariates).

ANALYSIS OF VARIANCE (ANOVA) An **inferential** statistical procedure used to test the **null hypothesis** that the **means** of three or more **populations** are equal to each other.

ALPHA LEVEL: In testing a **null hypothesis,** the alpha level is the probability of committing a **Type I error.** More explicitly, it is the probability of rejecting a true **null hypothesis.** Also called the "level of significance" of an hypothesis test.

ALTERNATIVE HYPOTHESIS: In **hypothesis testing,** the alternative hypothesis defines one or more values of the population **parameter** of interest. The alternative hypothesis typically includes a range of values of the population parameter, and is the converse of the **null hypothesis.**

BI-MODAL: A bi-modal **frequency distribution** is one that has two score values or two score classes that have higher frequencies than any other score values or score classes. A bi-modal **frequency polygon** would thus have two peaks (like a dromedary—or is it a camel?)

CHI SQUARE STATISTIC: This is one of many so-called "test statistics" that are used in **inferential statistics.** The chi-square statistic is used in tests of the statistical independence of two variables, and in tests of the identity of two **frequency distributions.** Depending on the value of a parameter called its "degrees of freedom," the chi-square statistic is compared to one of a family of known probability distributions, called the chi-square distributions. p. 212

CONCURRENT VALIDITY: A type of **validity** that represents the degree of correspondence between a new measurement instrument and a well-established, independently-validated measure of the same **construct.** Concurrent validity is usually expressed in terms of a **correlation coefficient** between scores on the new measurement instrument and scores on the well-established measurement instrument.

CONFIDENCE INTERVAL: A confidence interval consists of two values (a lower confidence limit and an upper confidence limit) that provide estimated bounds on a population **parameter.** A confidence interval is the result of interval **estimation.** When a researcher claims to be "95 percent confident that a population **parameter** falls within a given confidence interval," that person is stating a level of belief that the confidence interval contains the true value of the population **parameter.** The level of belief is not a probability value.

CONSTANT: A constant is any identical characteristic of all members of a **sample** or **population.** The sex of all members of the population of adult males in the United States is a constant. The converse of a constant is a **variable;** a characteristic on which members of a **sample** or **population** differ from each other.

CONSTRUCT: A construct is an unobservable, constructed variable that is used to label a consistent set of behaviors or observable variables. Many of the most important variables in education and the social and behavioral sciences are constructs. Examples include intelligence, socio-economic status, motivation, social class, anxiety, and achievement. Because they are not observable, constructs cannot be measured directly. A major task of social and behavioral research is the development of measures of various constructs.

CONSTRUCT VALIDITY: A type of **validity** that indicates the degree to which an instrument is a measure of the **construct** that it purports to measure. Some researchers assert that construct validity is the essence of all validity. Construct validity is never proved conclusively. To support a claim to construct validity requires a series of empirical demonstrations in the context of a well-defined theory.

CONTENT VALIDITY: A type of **validity** that is concerned with the correspondence between the content of a measurement instrument and the domain of knowledge it purports to represent. Content validity is determined judgmentally, not by using a statistical procedure. To be content valid, a measurement instrument should assess a representative sample of the knowledge or behavior domain which it is claimed to measure.

CONTINGENCY TABLE: A two-dimensional table that is defined by the levels of two categorical variables. The rows of the table are defined by the levels of one variable, and the columns of the table are defined by the levels of another variable. The cells of the table (each cell is formed by the intersection of a row and a column) contain the numbers of observations that are at the respective combinations of levels of each variable. For example, a contingency table would be formed by classifying a sample of adults by sex and political party affiliation (Republican, Democrat, or Independent). The resulting table would have two rows (male and female), and three columns (one for each political party). Each of the six cells of the table would contain the number of adults of that respective sex and party affiliation.

CORRELATION COEFFICIENT: This is a **sample statistic** or **population parameter** that indicates the degree of relationship between two variables. There are many kinds of correlation coefficients. The most frequently used are the **Pearson Correlation Coefficient,** a measure of the degree of linear relationship between two variables that are measured on an **interval scale,** and the **Spearman Correlation Coefficient,** a measure of the strength of linear relationship between two variables that are measured on an **ordinal scale.** Correlation coefficients can assume values between plus 1.0 and minus 1.0. A value of zero indicates no linear relationship; a value of plus 1.0 indicates a perfect direct relationship; and a value of minus 1.0 indicates a perfect inverse relationship.

CRITICAL REGION: In **hypothesis testing,** the critical region consists of the range of values of the test statistic that will lead to rejection of the **null hypothesis.** If the test **statistic** computed from **sample** data falls in the critical region, the **null hypothesis** is rejected; if the **test statistic** falls outside the critical region, the **null hypothesis** is retained.

CUMULATIVE FREQUENCY DISTRIBUTION: A cumulative frequency distribution is a two-column table that is used to summarize a set of data. One column contains an ordered list of the distinct score values in the data set. The other column lists the number of scores that are less than or equal to each corresponding score value. For example, if the data set consisted of the scores 1, 1, 2, 2, 2, 3, 4, the cumulative frequency distribution would be as follows:

Score Value	Cumulative Frequency
1	2
2	5
3	6
4	7

since two scores are less than or equal to 1; five scores are less than or equal to 2; etc.

DATA: This term describes the raw material of statistics. Data (plural; singular is "datum") are numbers that provide numerical values for any characteristic of a **sample** or a **population.**

DESCRIPTIVE STATISTICS: The branch of statistics that involves summarizing, tabulating, organizing, and graphing data for the purpose of describing a sample of objects or individuals that have been measured or observed. In descriptive statistics, no attempt is made to infer the characteristics of objects or individuals that have not been measured or observed.

ESTIMATION: A form of **inferential statistics** in which sample data are used to estimate the value of a population **parameter.** In one form, called "point estimation," the value of a sample **statistic** is used as a "best guess" of the value of a corresponding population **parameter.** In another form, called "interval estimation," sample statistics are used to form a **confidence interval,** a range of values within which a population **parameter** is thought to fall.

ESTIMATOR: An estimator is a formula that is used to compute an estimate of a population **parameter** from sample **data.** The formula for computing a sample **mean** is an example of an estimator.

F-TEST: An **hypothesis testing** procedure that is used for a variety of purposes. One application is a test of the **null hypothesis** that two **populations** have equal **variances.** Other applications are in the **analysis of variance** and the **analysis of covariance.**

FACTOR ANALYSIS: A statistical procedure that is used to reduce a large number of variables to a much smaller, representative set of variables, called "factors." The object of factor analysis is to achieve parsimony, and often to discover the essential variables that underlie and summarize the information in a large set of variables.

FREQUENCY POLYGON: A frequency polygon is a picture of a **grouped frequency distribution.** The horizontal axis (bottom) of a frequency polygon is marked off in midpoints of score intervals. The vertical axis (side) of a frequency polygon is marked off in frequencies or percent frequencies. The graph is constructed by plotting points that correspond to the frequencies or percent frequencies associated with the midpoint of each score interval. Adjacent points are then connected to form a graph. The graph consists of many connected straight lines, hence its name.

FREQUENCY DISTRIBUTION: A frequency distribution is a two-column table that is used to summarize a set of data. One column contains an ordered list of the distinct score values in the data set. The other column lists the number of times each score value appears in the original data set. These numbers are termed "frequencies."

GRAPH: A graph is a pictorial representation of a set of data. Popular graphs include **pie charts, histograms,** and **frequency polygons.** Just as a "picture is worth a thousand words," a good graph is worth at least 10,000 numbers in a table. Well constructed graphs provide a rapid view of the global features of a set of data, whereas tables often provide more-detailed information.

GROUPED SCORE DISTRIBUTION: A grouped score distribution is a two-column table that is used to summarize a set of data. The first column lists score intervals, score classes, or score groups (all synonymns). The second column lists the number of scores that fall into each score interval. Construction of grouped frequency distributions follows rules of experience, rather than theory. You will find similar, but not necessarily identical, instructions in any elementary statistics book.

HISTOGRAM: A histogram (commonly called a bar chart) is a picture of a **grouped frequency distribution.** The horizontal axis (bottom) of a histogram is marked off in limits of score intervals. The vertical axis of a histogram is marked off in frequencies or percent frequencies. The height of each bar in a histogram indicates the number or percent of cases in a data set that fall into its corresponding score interval.

HYPOTHESIS TESTING: A form of **inferential statistics** in which a decision is reached concerning the validity of an hypothesis about the value of a population **parameter.** The process proceeds in four steps: 1) A population **parameter** is hypothesized to equal some stated value; 2) data are collected on a sample of objects or individuals, randomly chosen from the population; 3) A test **statistic** is computed, using sample **data;** 4) a decision is reached to retain or reject the initial hypothesis, depending on the value of the test statistic.

INFERENTIAL STATISTICS: The branch of statistics that involves making inferences about the value of one or more population **parameters,** on the basis of sample **statistics.** The most common forms of inferential statistical procedures are **estimation** and **hypothesis testing.** Using the data gathered through a sample survey (such as the Gallup Poll) to estimate the proportion of voters who favor a political candidate is an example of statistical inference.

INTERACTION: When two independent variables interact in their effects on a dependent variable, the effect of one independent variable depends on the value of the other independent variable. For example, in an educational experiment, the effect of various curricula on student achievement might depend on the level of motivation of the students. In that case, curriculum and student motivation would be said to interact.

INTERVAL SCALE: This term labels one of four widely referenced sets of assumptions about the relationship between an underlying variable (often called a **construct**) and an observed variable, said to be a measure of the construct. If the construct is measured on an interval scale, a given interval on the scale of measurement corresponds to the same amount of the construct, regardless of where that interval occurs on the scale. The temperature on a thermometer that is calibrated in degrees Farenheit is an example of an interval scale, since a five degree increase on the scale is thought to measure the same temperature increase, whether it occurs between 20 degrees and 25 degrees or between 60 degrees and 65 degrees. The scale value zero is arbitrary on an interval scale, and does not represent absence of the construct. Note that zero degrees on the Farenheit scale, although a chilling number, does not represent total absence of temperature.

KURTOSIS: A statistic used to describe the peakedness of a **frequency distribution**. A distribution that is tall and thin is termed **leptokurtic;** a distribution that is about as peaked as a **normal distribution** is called **mesokurtic;** a distribution that is wide and flat, compared to a **normal distribution,** is termed **platykurtic.**

LEPTOKURTIC: A **frequency distribution** that is leptokurtic is tall and thin in shape. Any distribution that is markedly more peaked than the **normal distribution** is termed leptokurtic.

LEVEL OF MEASUREMENT: This term refers to the relationship that is assumed to exist between an underlying variable (often called a **construct**) and an observed variable, thought to be a measure of the construct. The four most widely referenced sets of assumptions about such relationships are termed **nominal**-level measurement, **ordinal**-level measurement, **interval**-level measurement and **ratio**-level measurement.

MEAN: This sample **statistic** or population **parameter** is the most frequently used indicator of the middle of a score distribution. Its other names are the "average" and the "arithmetic average." For a group of **raw scores,** it is found simply by summing the scores and dividing by the number of scores.

MEASURES OF ASSOCIATION: This generic term includes all **sample statistics** and **population parameters** that indicate the degree of relationship between two **variables.** Examples include the **Pearson Correlation Coefficient,** the **Spearman Correlation Coefficient** and the **chi-square statistics.**

MEASURES OF CENTRAL TENDENCY: This generic term includes all **sample statistics** and **population parameters** that indicate where the middle of a score distribution falls. The most common measures of central tendency are the **mean** (or arithmetic average), the **median**, and the **mode**. These **statistics** and **parameters** tell us where the center of a distribution "tends to be."

MEASURES OF VARIABILITY: This generic term includes all **sample statistics** and **population parameters** that indicate the spread of a score distribution. Examples include the **range**, the **standard deviation**, the **variance**, and the **semi-interquartile range**.

MEDIAN: This **sample statistic** or **population parameter** is an indicator of the middle of a score distribution. It is equal to the point on a score scale that divides a distribution exactly in half, in the sense that half the scores fall above the median and half the scores fall below. Other names for the median are the **50th percentile** and the **second quartile**.

MESOKURTIC: A **frequency distribution** that is mesokurtic is neither tall and skinny nor wide and flat; it is about as peaked as a **normal distribution**. In fact, the **normal distribution** is used as a standard for describing **kurtosis**, which means "peakedness."

MODE: The mode is sometimes an indicator of the middle of a score distribution, and is therefore classified as a **measure of central tendency**. For a group of scores, the mode is defined as the score value that occurs more often than any other score value. However, all scores in a group might occur with the same frequency (in which case there would be no mode), or a group of scores might have two score values that occur with the same highest frequency (e.g., there might be ten 5's and ten 8's in a **sample** of scores where there are no more than five occurrences of any other score value). In this case there would be two modes. Finally, even though a score distribution has only one score that occurs more frequently than any other, that score value might not fall in the middle of the distribution (think of the distribution of heights of a group composed of five basketball players and 50 typical 20-year-old males). The principal advantage of the mode is that it only requires **nominal-level** measurement.

MULTIPLE CORRELATION COEFFICIENT: This sample **statistic** or population **parameter** summarizes the degree of linear relationship between a criterion variable (a variable that is predicted or dependent) and a linear combination of independent variables (used as predictors). The multiple correlation coefficient tells a researcher how well a given variable can be predicted using a number of other variables as predictors. It is used in conjunction with **multiple regression** analysis.

MULTIPLE REGRESSION: A statistical procedure that is used to predict the value of one variable (called a dependent variable) from the values of at least two other variables (called independent variables). A procedure called the "least squares method" is used to find the linear combination of independent variables that does the best job of predicting the dependent variable.

NEGATIVELY SKEWED: A **frequency distribution** that is negatively skewed is shaped so that the majority of its scores fall above its **mean**. The distribution of heights of professional basketball players is negatively skewed, since most professional basketball players are quite tall, but a few of them are relatively short.

NOMINAL SCALE: This term labels one of four widely referenced sets of assumptions about the relationship between an underlying variable (often called a **construct**) and an observed variable, said to be a measure of the construct. If the observed variable measures the construct on a nominal scale, it is not assumed that a larger scale value corresponds to a larger amount of the construct. Nominal scale values merely serve as labels for different categories of the construct, and have no other meaning in terms of amount or quantity. The following numerical scale could be adopted to distinguish between people with different eye color: 1 if brown eyes, 2 if blue eyes, 3 if black eyes, and 4 if green eyes. A person categorized as a 4 would have a different eye color than a person categorized as a 2, but would not have "more" eye color.

NORMAL DISTRIBUTION: The normal distribution is a theoretical contrivance that is the backbone of **inferential statistics.** Without it, statisticians would be lost. A graph of a normal distribution is shaped like a bell. In fact, the normal distribution is often called the "bell-shaped curve." The normal distribution is often a good approximation to many actual **frequency distributions.** Statisticians have proven that **frequency distributions** of **sample means** almost always look like normal distributions.

NULL HYPOTHESIS: In **hypothesis testing,** the null hypothesis is an initial statement of belief about the value of a population **parameter.** The object of **hypothesis testing** is to reach a decision on the truth or falsity of the null hypothesis that the population **parameter** of interest is equal to a specified value.

ORDINAL SCALE: This term labels one of four widely referenced sets of assumptions about the relationship between an underlying variable (often called a **construct**) and an observed variable, said to be a measure of the construct. If the observed variable measures the construct on an ordinal scale, a larger scale value is assumed to represent a larger amount of the construct. Data that are in **ranks** are usually assumed to be ordinally scaled, since a higher rank (smaller number) represents more of the construct than does a lower rank (larger number). If students are ranked in terms of their standing in a high school class, the student ranked 1 is thought to have a higher academic standing than the student ranked 2, etc. However, it is not assumed that the difference in academic standing between the student ranked 1 and the student ranked 2 is the same as the difference between the student ranked 2 and the student ranked 3.

PARAMETER: A parameter is a summary value that describes some characteristic of a **population.** Examples of parameters include **means, standard deviations, proportions,** and **correlation coefficients.** The correlation between height and weight for the **population** of adult women in the United States is a parameter.

PEARSON CORRELATION COEFFICIENT: A sample **statistic** or population **parameter** that indicates the degree of linear relationship between two **variables** that are measured on an **interval scale.** The Pearson correlation coefficient can assume values between plus 1.0 and minus 1.0. A value of plus 1.0 indicates a perfect direct linear relationship; a value of minus 1.0 indicates a perfect inverse linear relationship.

PERCENTILE: A percentile is a point on a score scale that divides a score distribution into two parts. There are 99 different percentiles—the 1st percentile, the 2nd percentile, . . . up to the 99th percentile. Together the percentiles divide a score distribution into 100 parts. For example, the 25th percentile is the point on the score scale where exactly 25 percent of the scores in a distribution either equal it or fall below it.

PERCENTILE RANK: The percentile rank associated with a given **raw score** is the percentage of scores in a distribution that are less than or equal to that raw score. Percentile ranks are used to interpret an individual's **raw score** by comparing it with the performances of others measured with that individual. For example, if a raw score of 68 on a nationally normed test had a corresponding percentile rank of 55, one could conclude that 55 percent of the individuals in the norms sample earned **raw scores** of 68 or less.

PIE CHART: A pie chart is a type of graph used to display the percent of cases in a data set that fall into various categories. The graph is called a pie chart because it is circular in shape, with a wedge of the circle identified with each category. The width of each wedge is proportional to the percent of cases that fall into its corresponding category. Thus a category that contained 10 percent of all cases would have a wedge that was twice as wide as the wedge of a category that contained only 5 percent of all cases.

PLATYKURTIC: A **frequency distribution** that is platykurtic is wide and flat in shape. Any distribution that is markedly less peaked than the **normal distribution** is termed platykurtic.

POPULATION: Formally, a population is any collection of objects or individuals that have at least one characteristic in common. In **inferential statistics** one usually wishes to determine the value of some characteristic of a population, called a population **parameter**. Populations are the objects of generalization in **inferential statistics**.

PREDICTIVE VALIDITY: A type of **validity** that is represented by the effectiveness with which a measurement instrument can be used to predict future performance on some criterion variable. Predictive validity is usually summarized in terms of a **correlation coefficient** between scores on the measurement instrument and scores on a measure of the criterion variable. For example, the predictive validity of a test of reading readiness administered to beginning first-graders, might be measured in terms of its **correlation** with those students' second-grade reading achievement scores.

POSITIVELY SKEWED: In a positively skewed **frequency distribution,** a majority of the scores fall below the **mean.** The distribution of scores of sixth-grade students on a beginning calculus test would be positively skewed, since most sixth-graders would earn low scores, but a few (who learned some calculus from older siblings, from their parents, or on their own) would do quite well.

POWER: In **hypothesis testing,** the power of a test is the probability that a false **null hypothesis** will be rejected, in favor of a true **alternative hypothesis.** Power is equal to the converse of the probability of committing a **Type II error.**

RANGE: The range is a measure of the spread of scores in a distribution. It is most often used as a **descriptive statistic.** The range is equal to the difference between the largest and the smallest scores in a distribution.

RANKS: Ranks are used to convert a set of data to an **ordinal scale.** The original data are first listed from largest to smallest. The largest score is assigned a rank of 1; the next largest a rank of 2; etc. If two scores have the same value, they are given the average of the ranks they would have been assigned, had their values been different. For example, if the data set consisted of the scores 10, 9, 9, and 8, the corresponding ranks would be 1, 2.5, 2.5, and 4.

RATIO SCALE: This term labels one of four widely referenced sets of assumptions about the relationship between an underlying variable (often called a **construct**) and an observed variable said to be a measure of the construct. If the observed variable measures the construct on a ratio scale, a given interval on the scale of measurement coresponds to the same amount of the construct, regardless of where that interval occurs on the scale. In addition, zero on the scale of measurement corresponds to complete absence of the construct. Height measured in inches is an example of measurement on a ratio scale. The four inches between 48 inches and 52 inches represent the same difference in height as does the four inches between 58 inches and 62 inches. Also, a person who measured zero inches from the bottom of their feet to the top of their head would have no height at all. (I've never seen a person who was that short.)

RAW SCORE: A piece of data that has not yet been cooked is called a raw score. To be "raw" a score cannot be transformed to a derived scale, such as a **percentile rank** or a **standard score**.

RELIABILITY: A measurement concept that represents the consistency with which an instrument measures a given performance or behavior. A measurement instrument that is reliable will provide consistent results when a given individual is measured repeatedly under near-identical conditions. For example, a reliable bathroom scale would produce very similar weight measurements if a given individual were to step on the scale, then off, then on, then off, etc. Reliability is usually represented by a **correlation coefficient,** and can assume values between zero and 1.0. In practice, perfectly reliable measurement instruments are never found, but some standardized achievement tests have reliabilities around 0.95.

SAMPLE: A sample is any subset of a **population**. In **inferential statistics,** one usually attempts to select a sample that represents a **population,** and then collect data on all members of the sample. Sample **statistics** are then computed, in order to infer the value of population **parameters**.

SAMPLING DISTRIBUTION: A sampling distribution is the frequency distribution that a sample **statistic** or **estimator** would follow, if a researcher were to select repeated, independent samples and compute a value of the **statistic** or **estimator** for each sample. Sampling distributions are essential to **inferential statistics,** since they allow a researcher to determine the likelihood that **statistics** or **estimators** of a given magnitude will be observed.

SCATTER DIAGRAM: A scatter diagram is a type of **graph** that is used in conjunction with **correlation coefficients.** Each plotted point in a scatter diagram represents the values of an individual observation on the two **variables** being correlated. Values of one variable are shown on the horizontal axis (bottom) and values of the other are shown on the vertical axis (side of the graph).

SEMI-INTERQUARTILE RANGE: This **sample statistic** or **population parameter** is a measure of the spread of scores in a distribution. The more spread out the scores, the larger the semi-interquartile range. The semi-interquartile range is not used as often as the **standard deviation,** but has the advantage that it only requires **ordinal-level** measurement. It is equal to half the distance on the score scale, between the 75th percentile and the 25th percentile.

SKEWED: A skewed **frequency distribution** is not **symmetric** in shape. If more than half the scores in a distribution fall below its **mean,** the distribution is called **positively skewed;** if more than half the scores in a distribution fall above its **mean,** the distribution is called **negatively** skewed.

SPEARMAN CORRELATION COEFFICIENT: The Spearman correlation coefficient is a sample **statistic** or population **parameter** that indicates the degree of linear relationship between two **variables** that are measured on an ordinal scale. The Spearman coefficient provides the correlation between the ranks of two **variables.** It can assume values between plus 1.0 and minus 1.0.

STANDARD DEVIATION: This **sample statistic** or **population parameter** is the most popular measure of the spread of scores in a distribution. The more spread out the scores, the larger the standard deviation. The standard deviation requires **interval-level** measurement, and is equal to the square root of the **variance;** a distribution with a **variance** of 16 would thus have a standard deviation of 4.

STANDARD SCORE: A standard score is a type of derived score that is the result of transforming a variable to a scale having a specified **mean** and **standard deviation.** Widely used standard score scales include the z-score, which has a **mean** of zero and a **standard deviation** of 1.0; the Z-score, which has a **mean** of 50 and a **standard deviation** of 10; and the T-score, which is scaled exactly like the Z-score, but is often transformed to produce a **normal distribution** of scores.

STATISTIC: A statistic is a number that provides a summary of some characteristic of a set of data. Statistics are usually associated with **samples,** rather than **populations.** Examples of statistics include **means, ranges, correlation coefficients** and **standard deviations.** The average height of a group of fourth-grade students is a statistic. So is the proportion of Fortune 500 executives having lunch at the Ritz on Monday who put on their pants one leg at a time.

SYMMETRIC: A **frequency distribution** that is symmetric has the same shape, both above and below its **mean.** In fact, if you were to fold a symmetric **frequency polygon** in half, around its **mean,** its lower half would fall right on top of its upper half. In a symmetric **frequency distribution,** half the scores fall below the **mean** and half the scores fall above it.

t-TEST: Several widely used **hypothesis testing** procedures are termed "t-tests." One such procedure is used to test the **null hypothesis** that a **population mean** is equal to a specified value. Another t-test is used to examine the null hypothesis that the **means** of two **populations** are equal to each other.

TYPE I ERROR: In **hypothesis testing,** a Type I error is committed when a true **null hypothesis** is erroneously rejected. The probability of committing a Type I error is usually denoted by the Greek letter alpha.

TYPE II ERROR: In **hypothesis testing,** a Type II error is commited when a false **null hypothesis** is erroneously retained. The probability of committting a Type II error is usually denoted by the Greek letter beta.

VALIDITY: A measurement concept that is concerned with the degree to which a measurement instrument actually measures what it purports to measure. Validity is not absolute, but depends on the context in which a measurement instrument is used. It has been noted that no measurement instrument has inherent validity. One can make both valid and invalid inferences from the results of measurement. A number of types of validity have been identified in the measurement literature. They include **content validity, construct validity, con-**

current validity and **predictive validity.**

VARIABLE: A variable is any characteristic on which the elements of a **sample** or **population** differ from each other. Height and weight are variables, as are sex and national origin group. The converse of a variable is a **constant,** which is an identical characteristic for all members of a **sample** or **population.**

VARIANCE: This **sample statistic** or **population parameter** is an indicator of the spread of scores in a distribution. The more spread out the scores, the larger the variance. The variance is used quite frequently in **inferential statistics,** but less often in **descriptive statistics.** The variance is the square of the **standard deviation;** a distribution that had a **standard deviation** of 4 would thus have a variance of 16.

Answers to Problems

Chapter 1

1.1 The correct answer is b., since 75 out of 100 students scored below 36. Four students scored between 46 and 50, so a. is not correct. Since the majority of students scored below 36, c. is not correct. Two students scored at or below 10, so d. is not correct.

1.2 The correct answer is c., since the distribution has two "humps" and the majority of the distribution falls in the low end of the score scale. The distribution is not symmetric and, having more than one peak, it is not unimodal.

1.3 The category "6-10" has the tallest bar, and therefore has the highest frequency of schools associated with it. The frequency of that category can be read from the vertical scale of the histogram as 8.

1.4 Response a. is true, since there were 1,882 third-grade participants and 1,117 first-grade participants. Response b. is false, since there were 1,707 second-grade participants. Response c. is true, since a standard score of 50 is the national average, and 52.6 is the average standard score on the post-test for first-graders. This is higher than the average post-test standard score for second-graders or third-graders. Response d. is true, since the average gain for first-graders was five points (from 47.6 to 52.6), and this is higher than the corresponding gains in other grades.

Chapter 2

2.1 The correct answer is d. The average wouldn't be good for projection purposes, because the $210,000 spent in one year would inflate the projection too much. The mode wouldn't be useful for several reasons. First, there might not be one. Second, the mode ignores most of the data and is therefore not a good predictor of future performance. The smallest of the annual expenditures would undoubtedly underpredict future expenditures.

2.2 Response a. is correct. Whenever a distribution is unimodal, and the mean is smallest, the median is between the mean and the mode, and the mode is largest, the distribution is negatively skewed. The problem did not state that the distribution was unimodal, but it's a safe bet since we're considering a distribution of test scores for 40,000 students.

2.3   Response a. is true, since in a positively skewed distribution, more than half of the distribution falls below the mean. In this case, you know that is true, since the median is given as $2.10. Response b. is true, since the median of $2.10 divides the payoff distribution exactly in half. Response c. is false, since the median is $2.10, not $2.00. Response d. is false, since the mode is $2.00, not $3.75.

2.4   a. In all likelihood, the average is not an appropriate statistic to use in summarizing the morale data because the scale probably measures morale at the ordinal level of measurement, and not at the interval level.

b. You should not accept the conclusion that morale in Department B is more than twice as high as that of employees in Department A. To support that conclusion, you would have to believe that the scale measured employee morale at the ratio level of measurement, which is very unlikely. If ratio level measurement held, an employee who scored zero on the measurement scale would have no morale at all.

c. If you believe that the scale measured employee morale at least at the ordinal level of measurement, you should accept the conclusion that employees in Department B had a higher morale level than did employees in Department A.

Chapter 3

3.1   Response a. is false, since the polygon for girls is far wider than the polygon for boys. Response b. is true since the polygon for girls has far larger spread than the polygon for boys. Response c. is false, because response b. is true.

3.2   In a symmetric distribution, the middle 50 percent of the scores fall between the median plus the semi-interquartile range and the median minus the semi-interquartile range. In this case, the lower value is 28−4=24 and the upper value is 28+4=32.

3.3   Response a. is true, since five years is four standard deviations above the mean and most of a normal distribution is no more than three standard deviations above the mean. Response b. is true, since one year is four standard deviations below the mean and most of a normal distribution is no more than three standard deviations below the mean. Response c. is true because about 95 percent of a normal distribution falls between points that are two standard deviations below the mean (as is two years in this case) and two standard deviations above the mean (as is four years in this case). Response d. is false, because the median and the mean are equal (three years in this case) in a normal distribution.

3.4   a. The mean and standard deviation are not appropriate measures of central tendency and variability (respectively) in this case, since they both assume at least interval-level measurement. Rank-in-class is an ordinal-level variable.

b. Ordinal-level measurement is assumed when data are reported as ranks.

c. The median would be a more appropriate measure of central tendency, and the semi-interquartile range would be a more appropriate measure of variability.

Chapter 4

4.1   The correct response is d. since, in a general population that includes both children and adults, older people would tend to weigh more. Therefore the correlation between age and weight would be positive. However, it would not always be the case that an older person weighed more than a younger person, so the correlation would be less than 1.0.

4.2 Response a. is true, since a higher correlation, regardless of its sign, indicates a better predictor, and −0.85 is larger in magnitude than 0.60. Response b. is false, since it is the converse of Response a., and Response a. is true. Response c. is false, since Response a. is true. Response d. is true, since the correlation between number of absences and arithmetic achievement scores are negatively correlated.

4.3 Response a. is the appropriate conclusion, since correlation coefficients are always in the range −1.0 to +1.0. The other responses are not true, for a variety of reasons. Responses b. and d. are causal statements, which can never be supported solely on the basis of a correlation coefficient. Response c. ignores the obvious error in the reported correlation coefficient and is therefore false.

4.4 Response b. is true, since the scatter of points is upward and to the right (indicating a positive correlation) and the points do not fall right on a straight line (indicating a correlation coefficient that is less than 1.0).

Chapter 5

5.1 Response a. is correct because the test-retest method of estimating reliability indicates stability of measurement, whereas the KR-20 method only indicates the degree to which the items on the test tend to measure the same thing.

5.2 Response b. is correct because criterion-referenced tests are often designed to assess mastery of specific objectives, and a percentile rank is, inherently, a norm-referenced report of performance. A percentile rank specifies the performance of an individual examinee in terms of the performance of other examinees.

5.3 Response c. is correct because your objective is to use the test to select those who will do best in honors physics. Predictive validity is an indicator of the value of a measure in predicting future performance. Response a. is not correct because reliability is a necessary, but not a sufficient, component of high predictive validity. Response b. indicates high concurrent validity, which is not relevant for purposes of prediction. Response d. indicates high content validity, which is not relevant for purposes of prediction.

5.4 Response a. is correct, since a T-score of 70 is two standard deviations above the mean of the norm group, which corresponds to a percentile rank of about 97 in a normal distribution of scores; a percentile rank of 87 is well above average (which would be the 50th percentile in a symmetric distribution of scores); and a z-score of 0.25 is only one quarter of a standard deviation above the mean, which is pretty close to average performance. Based on this discussion, Responses b. and c. are clearly wrong. Relative to each other, Bob did best, Mary was next, and Joe did least well, so Response d. is wrong.

Chapter 6

6.1 Response c. is correct; the correlation coefficient would be a population parameter, since it was the object of estimation.

6.2 Response d. is correct. The 1500 voters constituted a sample selected from the population of potential voters.

6.3 Response b. is correct. The percentages are computed on the basis of sample data, and are therefore sample statistics. They could also be termed estimates of the population percentages that would be found, were all voters in the population to be polled.

6.4   Response a. is correct. A formula that is used to compute an estimate of a population parameter is called an estimator.

Chapter 7

7.1   Response b. is correct, since the 95 percent confidence interval is based on sample data. In fact, Response b. practically defines a 95 percent confidence interval. Response a. is not correct because, for a given randomly selected school system, the percentage of funding from state foundation programs is either between 43.2 and 67.8 (in which case the probability is 1.0) or it is not between those values (in which case the probability is zero). Responses c. and d. are not correct, because a confidence interval provides estimates of the value of a population parameter, not the true value of a parameter.

7.2   Response a. is correct, since 19.4 days is a sample statistic that provides a point estimate of the true population mean. Responses b., c., and d. are incorrect for a variety of reasons. The value 19.4 is not based on data from an entire population. Since 19.4 is a sample mean, it cannot be a lower confidence limit. Since the sample of classrooms was randomly selected, it is not the case that 19.4 has nothing to do with the true population mean; it is a good estimate.

7.3   a. This study is an example of inferential research. Undoubtedly, the meteorologists are interested in the relationship between sunspot activity and rainfall in future years, not just for the year that they sampled.

  b. The population would be all weeks—past, present and future.

  c. The sample would be the 52 weeks in the year that was observed.

  d. The +0.42 is a sample statistic, based on a sample of 52 weeks.

7.4   Response d. is correct, and is self-explanatory. Responses a., b., and c. are incorrect for a variety of reasons. No sample data can predict the future with certainty. Even if the Yankees had won all games in the past, one could not be sure that the Dodgers wouldn't make a comeback. Confidence intervals cannot be used to specify future probability values. Response c. is pure nonsense. Even if the upper confidence limit had been 1.00, the Dodgers would have some chance of winning in the future.

Chapter 8

8.1   Response a. is false for several reasons. First, because the data were not collected in a true experiment one cannot form causal conclusions. Second, testing a null hypothesis at a significance level of 0.05 does not provide "95 percent certainty." Response b. is false because the reported value is the outcome of a hypothesis test, and cannot be used to infer the width of a confidence interval. Response c. is true because the null hypothesis of no difference in population means was rejected with an alpha level of 0.05. Response d. is true, and is self-explanatory.

8.2   a. The authors might be generalizing to a nationwide population of elementary schools, or perhaps to a statewide population of such schools.

  b. The authors are testing the null hypothesis that, in the specified population of elementary schools, the correlation between age of school building and dropout rate equals zero.

  c. The alternative hypothesis implied by the authors' statement is that, in the specified population of elementary schools, there is a correlation between age of school building and dropout rate that is different from zero.

d. No, you cannot be certain that a correlation of +0.21 would be significant at the 0.05 level just because it is significant at the 0.10 level.

8.3 a. The alternative hypothesis in this study was that the percentage of movie-goers in the younger population was different from the percentage of movie-goers in the older population.

b. The null hypothesis would be rejected, because the test statistic fell into the critical region.

c. The test statistic was surely significant at the 0.05 level, because it was reported to be significant at the 0.01 level.

8.4 a. Null Hypothesis: For all students in the school system, the average motivation would be the same, regardless of which curriculum were to be used.

b. Alternative Hypothesis: The average motivation of students in the school system would be different, depending on which curriculum were to be used.

c. A Type I error would be committed if the null hypothesis of no difference in average motivation were to be rejected, even though both curricula were to produce the same average motivation.

d. A Type II error would be committed if the null hypothesis of no difference in average motivation were to be retained, even though one curriculum were to produce a higher average than the other.

Chapter 9

9.1 a. The null hypothesis was that the average sales produced by the population of radio advertising campaigns is equal to the average sales produced by the population of newspaper advertising campaigns.

b. The alternative hypothesis was that radio advertising campaigns and newspaper advertising campaigns are not equally effective, in terms of average sales produced.

c. The two asterisks attached to the t-statistic of 4.62 indicate, according to the footnote to the table, that the statistic was significant at an alpha level of 0.01. This means that the author tested a null hypothesis that the mean sales produced by radio advertising campaigns and newspaper advertising campaigns were equal, using a Type I error probability of 0.01, and that this null hypothesis was rejected.

9.2 A t-test for independent samples and a paired-difference t-test should be used with data collected in accordance with different research designs. The t-test for independent samples should be used when two statistically independent samples have been selected from a given population and then have been subjected, independently, to different treatments. It should also be used when independent samples have been selected from two different populations, and then have been subjected, independently, to the same or different treatments. The paired-difference t-test should be used when pairs of objects or persons have been randomly sampled from a single population, after which elements of the pairs have been subjected to two different treatments or have been measured independently. For example, the paired-difference t-test would be appropriate to test the null hypothesis that the mean IQ-test score of pairs of siblings is the same, even though the siblings have been raised in different environments.

Serious statistical errors can result when the paired-difference t-test is used inappropriately, when the t-test for independent samples should have been used instead. Likewise, use of the t-test for independent samples, when the paired-difference t-test should have been used, can result in serious statistical errors.

9.3 A non-directional alternative hypothesis is usually of the form: A population parameter is *not equal* to a specified value. In contrast, a directional alternative hypothesis is usually of the form: A population parameter is *greater than* a specified value, or, a population parameter is *less than* a specified value. The distinction has implications when the results of a research study are interpreted, and also has implications for the mechanics of a hypothesis test. When a non-directional alternative hypothesis is used (e.g., the population mean is not equal to 100), critical regions typically exist in both tails of the sampling distribution (e.g., reject the null hypothesis if the sample mean is either greater than 110 or less than 90). When a directional alternative hypothesis is used, a single critical region typically is defined in one tail of the sampling distribution (e.g., reject the null hypothesis if the sample mean is less than 95).

Chapter 10

10.1 a. The population to which the results of this study apply is presumably all adults in New York state. To support this generalization, it is necessary to assume that the 436 sampled adults were randomly sampled from all adults in New York state, and that some formal definition of an "adult" was used to determine who was and who was not a member of the population.

b. No, you cannot conclude on the basis of these data that additional formal education results in adults reading more books. This would be a causal interpretation, which cannot be supported solely on the basis of a correlation. It could be that adults complete more formal education and read more books because of some third variable that was not measured in the study.

c. A 95 percent confidence interval, computed using the same data, would have been wider than the 90 percent confidence interval reported in the study. If no additional data are collected, the effect of raising the level of confidence will always be to make the confidence interval wider.

d. Yes, the author could have used the same data to test the null hypothesis that years of formal schooling and number of books read per year were not correlated. The null hypothesis would have been that, in the population of adults in New York state, the correlation between years of formal schooling and number of books read per year equalled zero. Since the hypothesized value (zero) did not fall within the confidence interval, you can conclude that the null hypothesis would have been rejected at an alpha level of 0.10 (Note that this alpha level is 1.0 minus the level of confidence of the confidence interval).

10.2 a. The authors were testing the null hypothesis that teacher salaries and average student reading achievement were uncorrelated in the population of elementary school classes (grades one through five) in North Carolina.

b. The authors retained their null hypothesis, since they reported that the correlation was "not significant at the 0.10 level." One can therefore infer that the null hypothesis was tested using an alpha level of 0.10, and that the test statistic did not fall into a critical region.

c. These results might generalize to the population of all elementary school classes in grades one through five in the state of North Carolina. Such generalization would be based on the assumption that the researchers selected their 120 classes randomly, from a list of all such classes in the state.

d. The critical assumption that would be necessary to generalize these results to a larger population is that the sample is representative of the population. This would require the use of some probability sampling procedure (such as simple random sampling) in selecting the 120 classrooms used in the study.

10.3 a. The null hypotheses to which the footnote with one ''*'' refers are identical, except that one applies to a population of fourth-grade boys and the other applies to a population of fourth-grade girls. For each population, the null hypothesis states that the correlation between mathematics aptitude test score and mathematics achievement test score equals zero.

b. The null hypothesis to which the footnote with two ''**'' refers is that, for a population of fourth-grade boys and a population of fourth-grade girls, the correlations between mathematics aptitude test score and mathematics achievement test score are equal to each other.

c. It is difficult to say, based on the information given in the problem, what population might be appropriate for generalization of these results. If it were the case that the 88 fourth-grade girls and 85 fourth-grade boys had been randomly sampled from lists of fourth-graders in a single school system, then generalization to the populations of fourth-graders in the school system would appear to be warranted. The appropriate populations for generalization depend on how the fourth-graders were sampled for this study.

10.4 a. The research question you define should certainly involve the degree of linear relationship between two variables. If you intend to use a Pearson correlation coefficient in examining your research question, the variables used in your study should be measurable at least at the interval level of measurement.

b. The population parameter that is central to your study should be a correlation coefficient. The population you identify should be defined very carefully, so that you can clearly identify who or what is a member of your population, and who or what is not a member.

c. The null hypothesis that is relevant to your study might state that a population correlation coefficient is equal to zero, or that two population correlation coefficients are equal to each other. In the first case, the alternative hypothesis would be that the population correlation coefficient is not equal to zero. In the second case, the alternative hypothesis would be that the two population correlation coefficients were not equal to each other.

d. If you were to reject the null hypothesis, in the first case (see answer to part c.), you would conclude that the variables being correlated did have some degree of linear relationship in the population from which you had sampled. In the second case, you would conclude that the variables being correlated did not have the same degree of linear relationship in the populations from which you had sampled.

Chapter 11

11.1 Jefferson and Jackson were testing the null hypothesis that, in the population of voters of 1842, the two variables, political party affiliation and position on extension of the Voting Rights Act of 1834, were statistically independent.

11.2 Jefferson and Jackson were attempting to generalize their results to the population of voters of 1842 who were declared or registered as Republicans, Democrats, or Independents.

11.3 If they acted in accordance with their findings, Jefferson and Jackson rejected their null hypothesis. This conclusion is supported by the footnote to their table, that indicates the statistical significance of the chi-square statistic of 93.0, using an alpha level of 0.05.

11.4 a. The authors were testing the null hypothesis that scores on their standardized attitude scale and academic major were statistically independent variables, for some populations of students with majors in sociology, psychology, education, and political science.

b. The authors reached the decision to retain their null hypothesis, since they reported that the chi-square statistic was "not significant at the 0.10 level." Apparently, 0.10 was the alpha level they used in testing their null hypothesis, and the chi-square statistic of 7.04 did not exceed the critical value associated with this alpha level.

c. Interpretation of the results of this study depends on the assumptions you are willing to make about how the authors sampled the 50 students of each major they used in the study. Another assumption that is critical concerns the validity of the standardized attitude scale the authors used. Unless the scale is a valid measure of attitude toward education, any conclusion about the authors' results, either for their sample or for some larger population, would be suspect.

Chapter 12

12.1 The authors were testing the null hypothesis that, for the population of students registering for beginning statistics courses at the major university, the mean score on the final examination would be the same, regardless of the method used to teach statistics. The null hypothesis is therefore that all three teaching methods examined in the study are equally effective. The alternative hypothesis is that the three methods of teaching statistics are not equally effective; at least two of the methods would result in different mean scores on the final examination, were each method to be used with the entire population of students.

12.2 The authors rejected the null hypothesis, since the reported F-statistic (5.67) was larger than the reported critical value of 3.13. Also, the F-statistic was reported to be "significant at the 0.05 level" in the footnote of the table.

12.3 This study suggests that method of teaching beginning statistics makes a difference, in terms of average performance on a statistics final examination, and that this conclusion applies to a population of beginning students in a major university. Whether the results of this study have implications beyond the university where the research was conducted depends on several factors, including the characteristics of the sampled students, the precise nature of the instructional methods used, and the composition of the final examination used to measure the dependent variable.

12.4 a. Your research question should involve a comparison of three or more population averages.

b. The population parameters that are central to your research study should include three or more population averages.

c. Your null hypothesis should involve the equality of at least three population averages. Your alternative hypothesis should state that at least one pair of these population averages is different from each other.

d. If you were to reject the null hypothesis, you would have to conclude that at least one population had an average that was different from the average of at least one other population. Depending on the nature of your research, this conclusion might support the contention that at least one of a number of treatments was more effective than at least one other treatment.

Chapter 13

13.1 One null hypothesis was that the three methods of instruction were equally effective; i.e., that the population of beginning statistics students would have the same average score on the final examination, regardless of the method used. Another null hypothesis was that populations of low-aptitude and high-aptitude students would have the same average score on the statistics final examination. The third null hypothesis was that there were no interaction effects between instructional method and student aptitude; i.e., that the relative effectiveness of the three instructional methods was the same, regardless of student aptitude.

13.2 The researchers rejected all three null hypotheses, because all three F-statistics were described as being "significant" at either the 0.01 alpha level or the 0.05 alpha level (in the case of the interaction null hypothesis). The footnote to the table supports this conclusion.

13.3 Assuming that the results of this study generalize at least to the population of beginning statistics students at the major university, one can conclude that instructional method has an effect on performance on a statistics final examination, as does quantitative aptitude level. Furthermore, the relative effectiveness of the three methods of instruction examined in the study depends on the quantitative aptitude level of the students.

13.4 No, you cannot conclude that one of the instructional methods is more effective than the others for all students. Because the interaction effect was found to be significant, it might be the case that one instructional method is most effective for low-aptitude students, whereas another is most effective for high-aptitude students.

Chapter 14

14.1 a. The independent variable in this study was length of instruction per day or "engaged instructional time." The dependent variable was score on the Reading Comprehension subtest of the Iowa Tests of Basic Skills. The covariate was score on the verbal intelligence test administered at the beginning of the study.

c. The researchers were testing the null hypothesis that the population of third-graders would have the same average score on the Reading Comprehension subtest of the Iowa Tests of Basic Skills, regardless of the length of instruction in reading each day.

c. By using the analysis of covariance instead of the analysis of variance, Harbinger and Spring were able to eliminate possible differences in the achievement of the three samples, due solely to differences in their average verbal intelligence at the beginning of the study. The researchers were thus able to eliminate one potentially contaminating variable from their study, and gain statistical precision in their investigation of the effects of length of instruction on reading achievement.

14.2 a. Mother's education contributes most to the prediction of sixth-grade arithmetic achievement. This assertion is supported by two facts. First, mother's education was entered first in the stepwise multiple regression analysis. Second, the correlation of mother's education with the dependent variable is higher than the correlation of any other predictor with the dependent variable.

b. Twenty-nine percent of the variance of arithmetic achievement is predicted by the four independent variables. This assertion is supported by the "R-squared" value of .29 noted in the last line of the table. When .29 is multipled by 100 to yield a percent, the result is 29 percent.

    c. The null hypotheses the authors tested when they reported that their F-statistics were "significant at the 0.05 level" were all similar. At each step of the regression analysis, the authors were testing the null hypothesis that the variables then used as predictors (the independent variables in the regression equation) did not contribute to the prediction of arithmetic achievement. Statistically, they were testing the null hypothesis that the coefficient of determination (R-squared) was equal to zero at each step of the analysis.

14.3 In a factor analysis, a report that "three factors accounted for 76 percent of the variance of the ten items" means that three underlying variables (called factors) do a relatively good job of representing the relationships among ten original variables (items). Just over three-fourths of the variation among scores on the ten items is explainable by variation among scores on the three underlying factors. It must therefore be the case that the ten items cluster together, to a considerable degree, around three factors.